New Mastermat

Book 2

Paul Briten

OXFORD

OXFORD
UNIVERSITY PRESS

Great Clarendon Street, Oxford OX2 6DP

Oxford University Press is a department of the University of Oxford.
It furthers the University's objective of excellence in research,
scholarship, and education by publishing worldwide in

Oxford New York

Auckland Cape Town Dar es Salaam Hong Kong Karachi
Kuala Lumpur Madrid Melbourne Mexico City Nairobi
New Delhi Shanghai Taipei Toronto

With offices in

Argentina Austria Brazil Chile Czech Republic France
Greece Guatemala Hungary Italy Japan Poland Portugal
Singapore South Korea Switzerland Thailand Turkey Ukraine
Vietnam

First published 2004

British Library Cataloguing in Publication Data

Data available

ISBN 13: 978 019 8361138

ISBN 10: 0 19 836113 0

20 19 18

Acknowledgements
Autumn Header Image © istockphoto (www.istockphoto.com)

Illustrated by Martin Chatterton, Trevor Dunton, IFA Design Ltd.
Cover illustration by Jonatronix
Page make-up by IFA Design Ltd, Plymouth, UK
Printed in China by Leo Paper Products Ltd.

Contents

Introduction

The **Mastermaths** series will help you have fun practising all the maths you need to learn. Read this page first to see how to use the book.

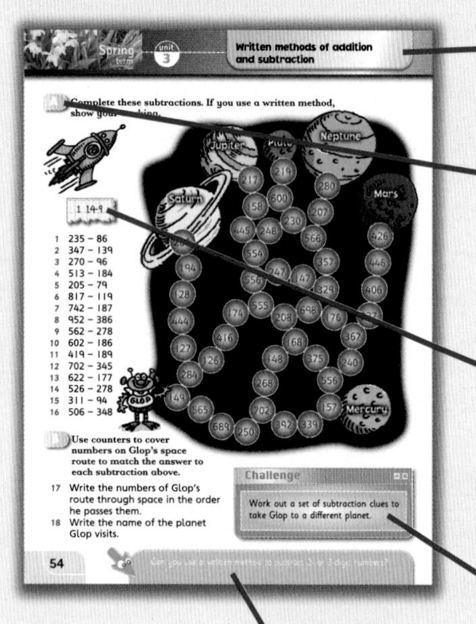

The title tells you what the page is about.

Work through the sections in order. Your teacher will tell you which questions to answer.

Some sections have examples to show how to set out your answer or remind you how to do the question. For sections with no example, you must choose your own method. Ask your teacher if you need help.

At the end of each page is a **Challenge**. Can you use the maths skills you have learned to solve the puzzles?

When you have finished working on a page, check that you can do this.

The **Think about it ...** pages are fun games and activities for you to try.

Check that you have everything you need before you start.

Read the instructions carefully to make sure you know what to do. Ask your teacher if you don't understand.

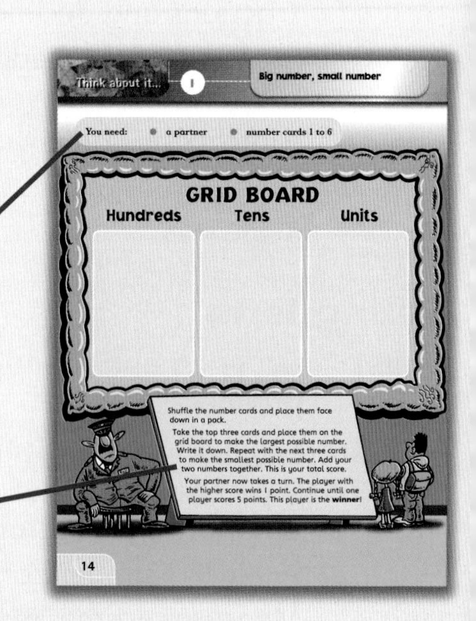

This book also contains six **Review** pages. Work through the questions in order to see if you have understood all the maths in the **Units**.

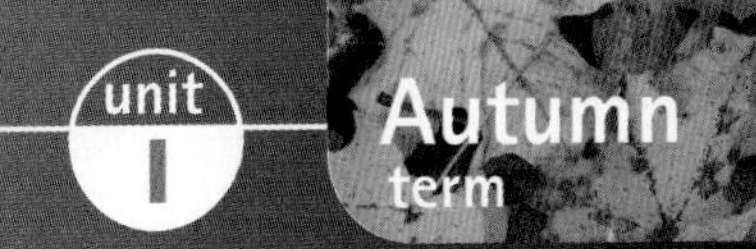

A Write in words.

1 426
2 247
3 328
4 460
5 708

1 four hundred and twenty-six

B Match the numbers to the words.

208 802 280

511 820

6 five hundred and eleven
7 eight hundred and two
8 two hundred and eighty
9 eight hundred and twenty
10 two hundred and eight

6 511

C Write the value of the green digits.

11 six hundred

11 621
12 567
13 402
14 819
15 741

D Write each set of numbers in order, smallest first.

16 36, 63, 639, 670, 936

16 639 63 936 36 670
17 17 171 71 117 711
18 326 236 623 263 632
19 443 343 433 434 334

E Write in words.

20 1964
21 4503
22 7026
23 10 000

F Write in figures.

24 four thousand two hundred and ninety-eight
25 two thousand one hundred and sixty-two
26 eight thousand seven hundred and seventy
27 six thousand and fifty-six
28 five thousand and nine

Challenge

Arrange these four digits to make:
a the largest number
b the smallest number
c five other numbers.
Write them in words.

6 4 1 7

Can you write 4-digit numbers in figures and in words?

A Break down these numbers.

1 3472
2 4276
3 1642
4 2019
5 9007

1 3000 + 400 + 70 + 2

B Write the missing numbers.

6 8265 = 8000 + 200 + □ + 5
7 6214 = 6000 + □ + 10 + 4
8 5707 = 5000 + □ + 7
9 7390 = □ + 300 + 90
10 4566 = 4000 + 500 + □ + 6

C Answer these.

11 2794 + 100
12 1365 + 10
13 1427 + 100
14 7368 − 10
15 9426 − 100
16 1234 + 1000
17 4628 − 100
18 6190 − 1000
19 4327 + 1000

D Round to the nearest 10.

20 67
21 24
22 16
23 94
24 66
25 21
26 85

20 70

E Round to the nearest 100.

27 6749
28 4671
29 9419
30 8365
31 1048
32 3850

27 6700

F Estimate the number reached by each snail.

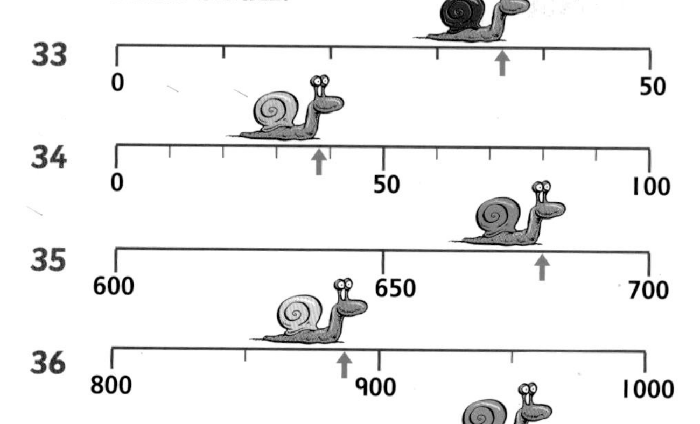

Challenge

Estimate how many bees in the hive.
Explain how you made your estimate.

Can you estimate the position of a point on a number line?

A

Write the temperature the thermometer shows on each day.
Write which day was:

1 −2 °C

6 warmest
7 coldest
8 1 °C colder than Monday
9 7 °C warmer than Tuesday
10 colder than Friday.

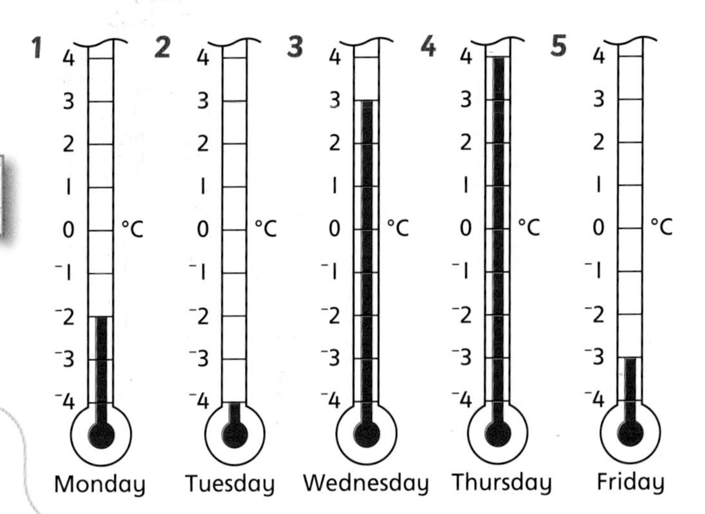

B

Write the four numbers missing from each number line.

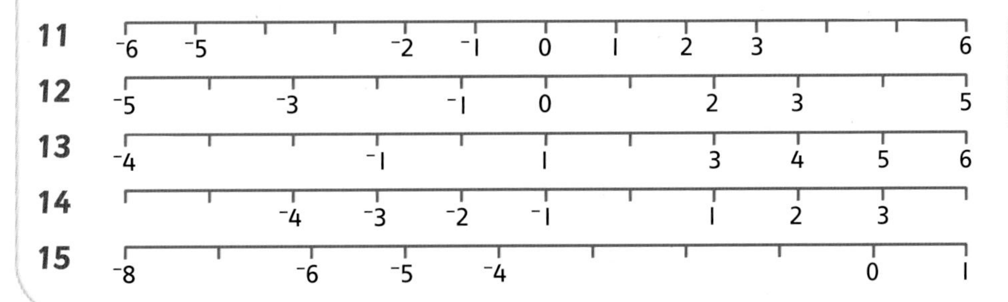

11 −4, −3, 4, 5

C

Write each set of temperatures in order, lowest first.

16 −8 °C, −3 °C, −2 °C, 4 °C, 7 °C

16 −2 °C 4 °C −8 °C −3 °C 7 °C
17 11 °C −10 °C −2 °C 4 °C 8 °C
18 −6 °C −7 °C 2 °C 0 °C −3 °C
19 1 °C −3 °C 13 °C −1 °C 2 °C
20 −1 °C −4 °C 8 °C 6 °C 0 °C

D

Write each new temperature.

21 from −4 °C rises by 5 °C
22 from −3 °C rises by 6 °C
23 from 3 °C falls by 3 °C
24 from −2 °C falls by 2 °C
25 from 4 °C falls by 7 °C

Challenge

Copy and complete this chain.

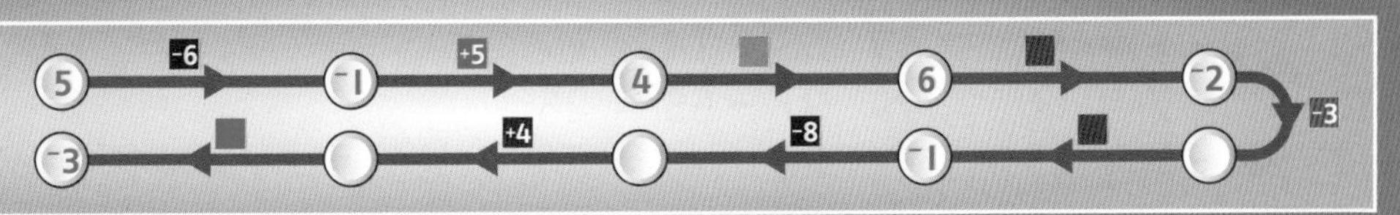

A

Use the code to find the words in this secret message.

1. 9 + 8, 16 + 4, 17 − 5
2. 14 + 5, 18 + 7, 19 − 14
3. 11 + 19, 14 − 9, 20 − 6, 15 + 14, 19 − 17, 7 + 13, 6 + 11, 13 + 16
4. 18 − 13, 17 + 13, 8 + 8, 16 − 4, 17 + 17
5. 5 + 7, 12 − 7, 19 + 18, 20 − 6, 13 + 8, 19 − 17, 12 + 18
6. 16 − 11, 18 + 14
7. 14 + 15, 14 + 11
8. 18 + 11, 13 + 18, 6 + 15, 3 + 9, 17 + 12, 12 + 7
9. Answer the question in code.

1 17, 20, 12 = C A N

B

Write two addition and two subtraction facts for each.

10 16 + 9 = 25
11 14 + 7 =21
12 28 − 7 = 21
13 26 + 25 = 51
14 60 − 17 = 43

10 16 + 9 = 25
9 + 16 = 25
25 − 16 = 9
25 − 9 = 16

C

Write an addition to check these subtractions.

15 19 − 6 = 13
16 18 − 7 = 11
17 22 − 7 = 15
18 43 − 28 = 15
19 68 − 19 = 49
20 91 − 65 = 26
21 80 − 16 = 64

D

Use these facts to mark Gary's homework. Correct his mistakes.

22 68 − 29 = 39

84 − 17 = 67
29 + 35 = 64
32 + 49 = 81
100 − 37 = 63
31 + 27 = 58
29 + 39 = 68

22 68 − 29 = 31
23 58 − 27 = 31
24 63 + 37 = 90
25 84 − 67 = 27
26 81 − 49 = 32
27 100 − 63 = 37
28 64 − 29 = 33

Challenge

Copy and complete these targets.

Can you check answers to subtraction questions by using addition?

A Write the missing information for each machine.

1 25

1

3

5

2

4

6

B Solve these word problems.

7 There are 58 people on a train. 19 get off. How many are left on the train?

8 A website receives 87 hits in the morning and 75 in the afternoon. How many hits are received altogether?

9 A class library has 102 books. If 37 are taken out, how many books are left in the library?

10 Between them, Ali and Sam have £47. If Ali has £29, how much has Sam?

C Complete these.

11 64 + 19

12 35 + 19

13 62 − 32

14 46 + 18

15 84 − 16

16 75 + 17

17 49 + 52

18 66 + 45

19 84 + 29

20 105 + 77

21 214 + 48

22 374 − 62

23 236 − 40

Challenge

Copy and complete these grids.

56	+ 19	75
39		16
	− 28	61
163		73

49		102
	+ 47	66
	− 27	127
256	−128	

Can you add and subtract by using multiples of 10 or 100?

A Draw the bubble, write the double.

1 26 52

1 26
2 16
3 27
4 44
5 35
6 79
7 131
8 180
9 260
10 341

B Use doubles to answer these questions. Show your working.

11 26 + 28
12 25 + 27
13 51 + 52
14 27 + 29
15 124 + 126
16 36 + 37
17 48 + 45
18 161 + 165

11 26 + 26 = 52
52 + 2 = 54
26 + 28 = 54

C Answer these.
Show your working.

19 536 + 49
20 86 + 75
21 94 + 38
22 324 + 217
23 489 + 68
24 75 + 879
25 2169 + 47
26 328 + 1468
27 5247 + 838
28 1645 + 366
29 2748 + 1662
30 9796 + 8927

Challenge

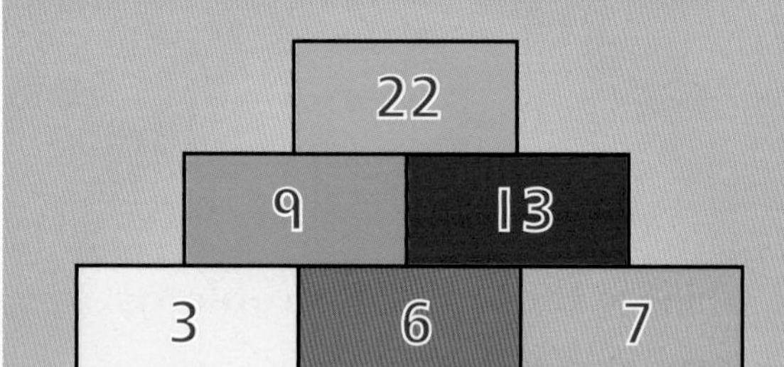

Work out how these number pyramids are formed.
Can you make a different number on the top row by changing the order of the numbers in the bottom row?
Design a pyramid with a top number of:

a 28
b 79
c 65
d 100

Can you use doubles to help you with addition?

A Find five children who are flying the right kite.

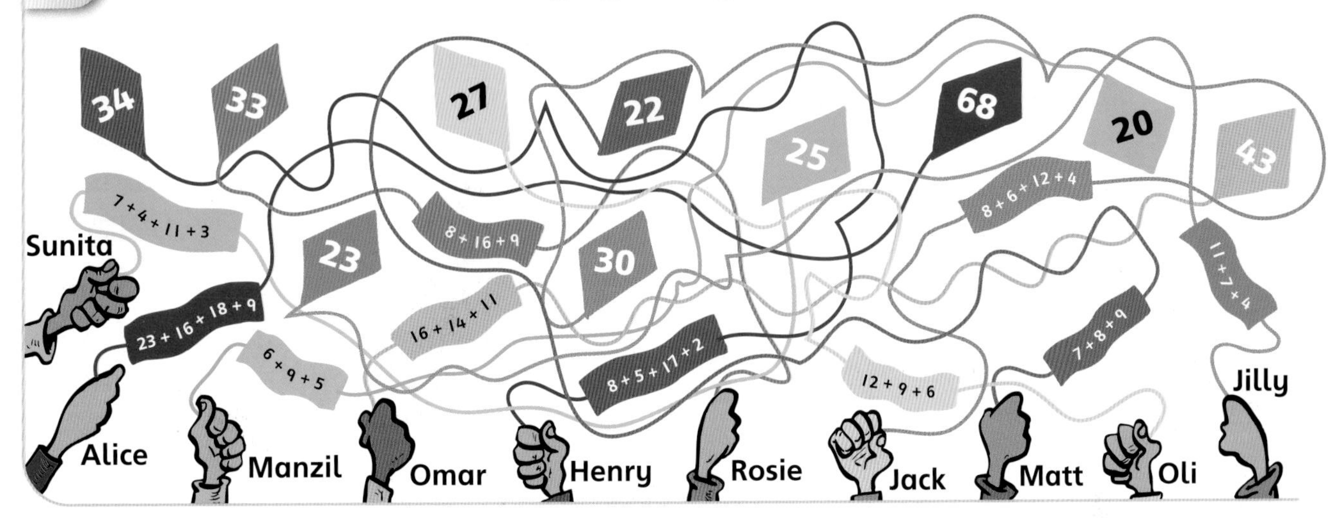

B Answer these questions.

6. 6 + 8 + 12
7. 9 + 7 + 11
8. 6 + 15 + 7 + 8
9. 18 + 14 + 12
10. 20 + 13 + 80 + 17
11. 7 + 8 + 7 + 6
12. 9 + 7 + 11 + 13
13. 70 + 80 + 30 + 40
14. 12 + 9 + 217
15. 115 + 19 + 118 + 17

C Write how much change from £1 for each.

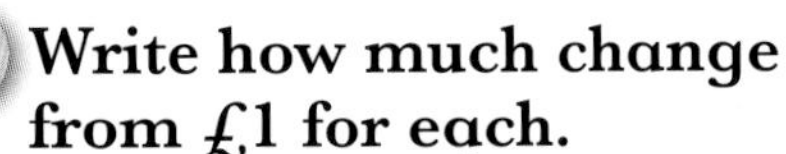

16. 1 rubber egg
17. 1 ink blot
18. 1 spider
19. 1 worm
20. 2 ink blots
21. 1 worm and 1 spider
22. 2 rubber eggs

Challenge

Plan a meal for two adults and two children. Write:

a the total cost of the four meals

b how much change from £100.

Menu

soup	£2	steak + chips	£12	ice-cream	£2
melon	£2	fish + chips	£8	trifle	£4
prawn cocktail	£4	lasagne	£6	apple pie	£4
		cheese salad	£8		

children's meals $\frac{1}{2}$ price

A Solve these money problems.

1 How much change from £1 when you buy a lemon?
2 How much change from £2 when you buy two cucumbers?
3 How much change from £5 when you buy three melons?
4 How many bunches of grapes can you buy with £10?
5 If you have £10, how much more money do you need to buy 4 kg of apples, two bunches of grapes and four melons?

B Write the difference between a first class and an economy ticket to each place.

	Flight to	1st class	Economy
6	TOKYO	£ 470	£ 389
7	SYDNEY	£ 643	£ 569
8	ALICANTE	£ 111	£ 88
9	NEW YORK	£ 627	£ 467
10	FIJI	£ 999	£ 853

C Write in pence.

11 164p

11 £1·64
12 £1·49
13 £2·62
14 £5·20
15 £0·63
16 £3·02
17 £5·51
18 £0·80
19 £7·03
20 £0·08
21 £0·10

D Write in pounds.

22 £0·58

22 58p
23 85p
24 162p
25 327p
26 463p
27 509p

Challenge

Use two counters and a dice. Place one counter on each track. Throw the dice. Move each counter by the number shown. Answer the question. How many different questions can you answer?

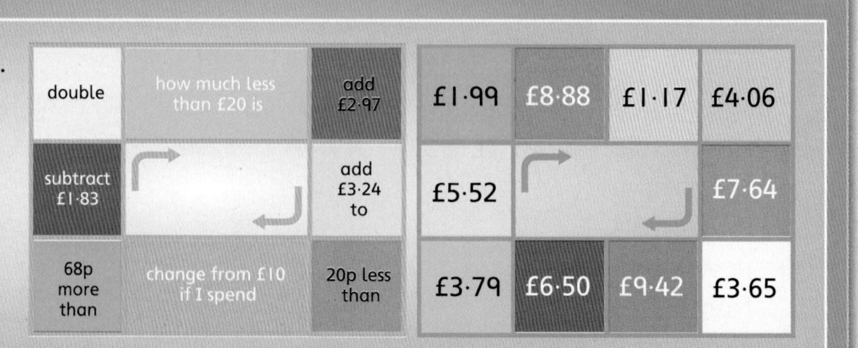

Can you add or subtract two sums of money?

A Solve these problems.

1. Penny Pincher spends £6·60, 57p and £4·44. How much does she spend altogether?
2. If you add 17 to a number the answer is 51. What is the number?
3. If you have £4·20 and spend £1·85, how much is left?
4. A farmer milks 135 cows in the morning and 98 in the afternoon. How many cows does he milk altogether?
5. If you have £10·60 and spend £3·90, will you have enough left to buy a toy costing £6·75?
6. The total of the ages of Grandad, Dad and Alex is 158. If Dad is 52 and Alex is 17, how old is Grandad?
7. There are 424 toffees in three jars. If there are 162 in the first jar and 95 in the second jar, how many are in the third jar?
8. Jesse and Billy together have 95p less than Ellie. If Jesse has £2·68 and Billy has £5·83, how much does Ellie have?

B Complete these additions. Show your working.

9. 423 + 296
10. 161 + 208
11. 372 + 423
12. 827 + 61
13. 49 + 827
14. 538 + 197
15. 365 + 566
16. 1264 + 1528
17. 3619 + 2190
18. 7382 + 1238
19. 4763 + 1828
20. 4618 + 4686
21. 5666 + 2948

C Complete these subtractions. Choose your method.

22. 263 − 149
23. 183 − 74
24. 375 − 168
25. 407 − 139
26. 5310 − 1185
27. 7903 − 289
28. 6720 − 4398
29. 9851 − 2587

Challenge

Work out how these rockets have been made. Write the missing numbers. Design a rocket with more sections than these.

Rocket 1:
3, 9, 6, 4, 8, 5
12, 15, 10, 12, 13
27, ☐, 22, ☐
52, ☐, 47
99, ☐
☐

Rocket 2:
115, 189, ☐, 272, 151, ☐
304, 361, 444, ☐, 576
665, 805, ☐, 999
1470, ☐, ☐
3142, ☐
☐

You need: a partner, number cards 1 to 6

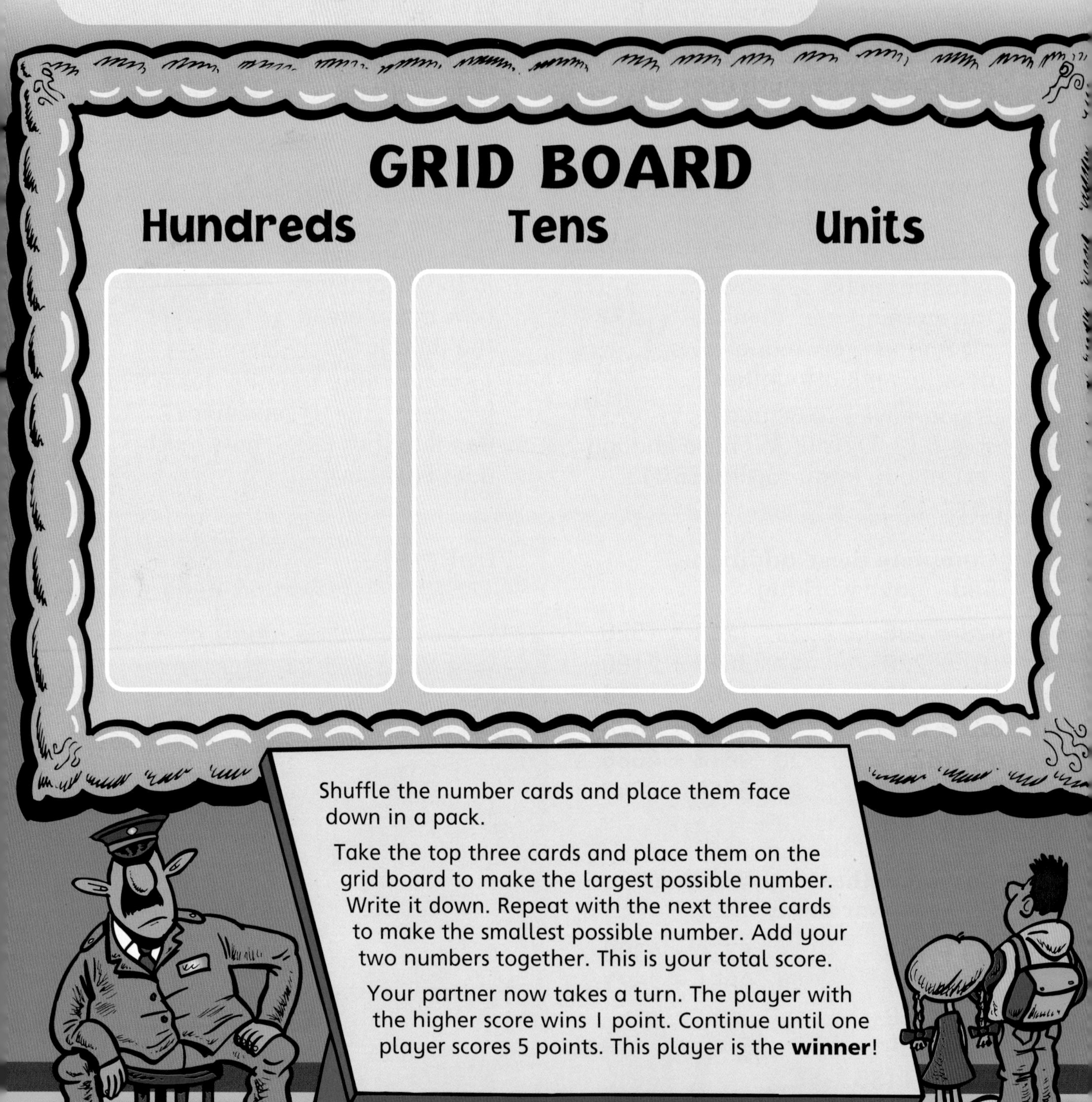

Shuffle the number cards and place them face down in a pack.

Take the top three cards and place them on the grid board to make the largest possible number. Write it down. Repeat with the next three cards to make the smallest possible number. Add your two numbers together. This is your total score.

Your partner now takes a turn. The player with the higher score wins 1 point. Continue until one player scores 5 points. This player is the **winner**!

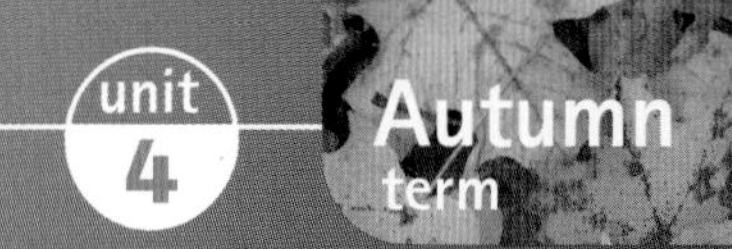

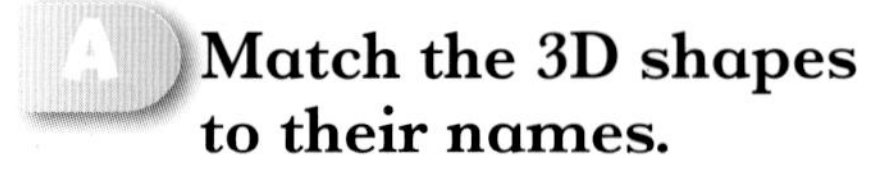

A Match the 3D shapes to their names.

1 2 3 4 5 6 7

triangular prism	pyramid	hemisphere	hexagonal prism	rectangular prism	cylinder	cone

B Write how many:

8 faces on a triangular prism
9 faces on a rectangular prism
10 edges on a hexagonal prism
11 vertices on a triangular prism.

C Look at these shapes packed in boxes. Write what different shapes each could be.

12 It could be a cylinder, or a ...

12

13

14

D Write the name of each shape. Explain your answers.

15 It is a rectangle because it has four right angles and the opposite sides are equal.

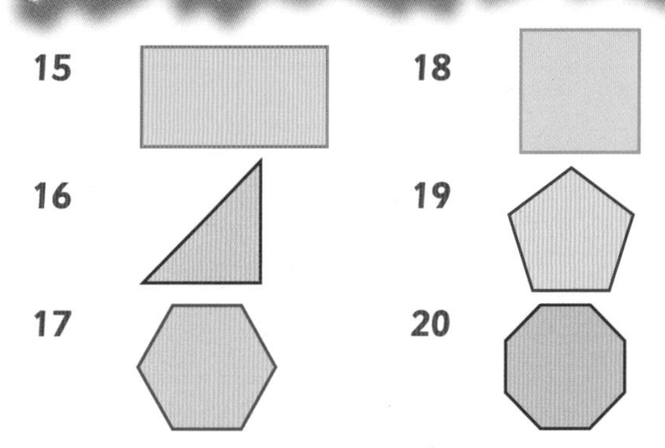

Challenge

Copy this Carroll diagram. Write the names of as many polygons as you can in the diagram.

	1 or more right angles	no right angles
quadrilateral		
not a quadrilateral		regular pentagon

A Answer these questions.

1. What is the name of this 3D shape?
2. How many edges does it have?
3. How many faces does it have?
4. How many vertices does it have?
5. What is the shape of each face?

B Copy and complete the table.

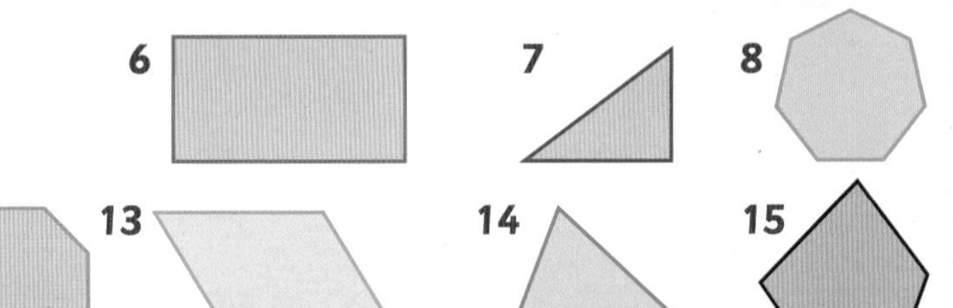

9 10 11 12 13 14 15

Polygon	Name of polygon	Number of right angles	Does it have line symmetry?	Is it regular?
6	rectangle	4	✓	✗
7				
8				
9				
10				
11				
12				
13				
14				
15				

Challenge

Investigate several regular polygons. Give the name of each polygon. Find how many sides and lines of symmetry each has. What do you notice?

Can you describe and visualize 3D and 2D shapes including the tetrahedron and the heptagon?

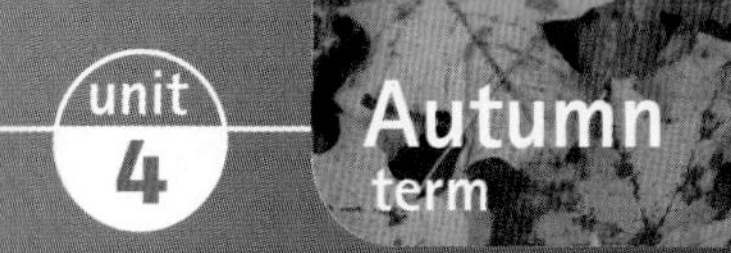

A

Part of each polygon is hidden. Write two shapes each could be and two each could not be.

> 1 could be: rectangle, pentagon
> could not be: triangle, square

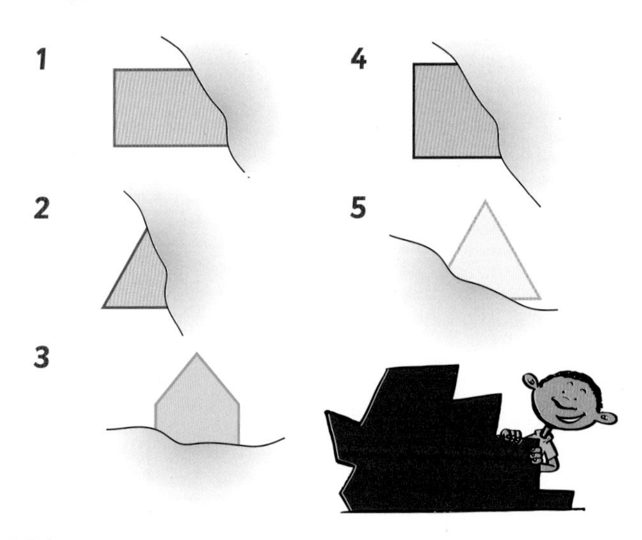

B

Write the correct name for each triangle: isosceles, equilateral or scalene.

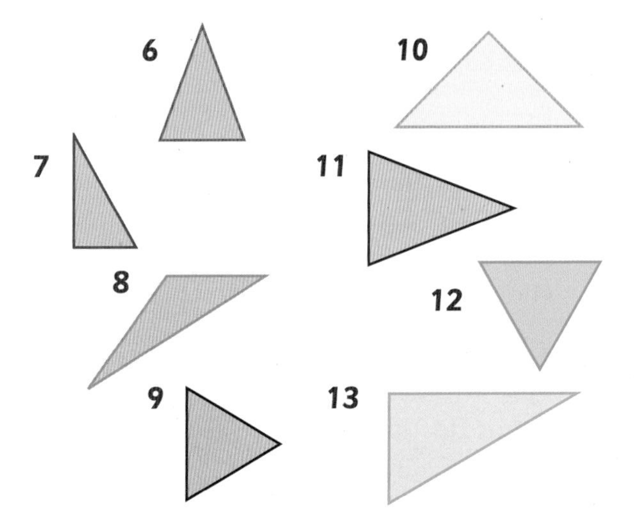

Tick those that are right-angled.

C

Name these shapes.

14 a 3-sided polygon with two equal sides and no right angles

15 a quadrilateral with four right angles and four equal sides

16 a polygon with three sides, all sides of different lengths

17 a quadrilateral with four equal sides and no right angles

18 a 5-sided polygon with all sides the same length

19 a quadrilateral with two sides of different lengths and all its angles equal

Challenge

Use six equilateral triangles of the same size. Using any number of triangles make as many different 2D shapes as you can. Draw and name your shapes.

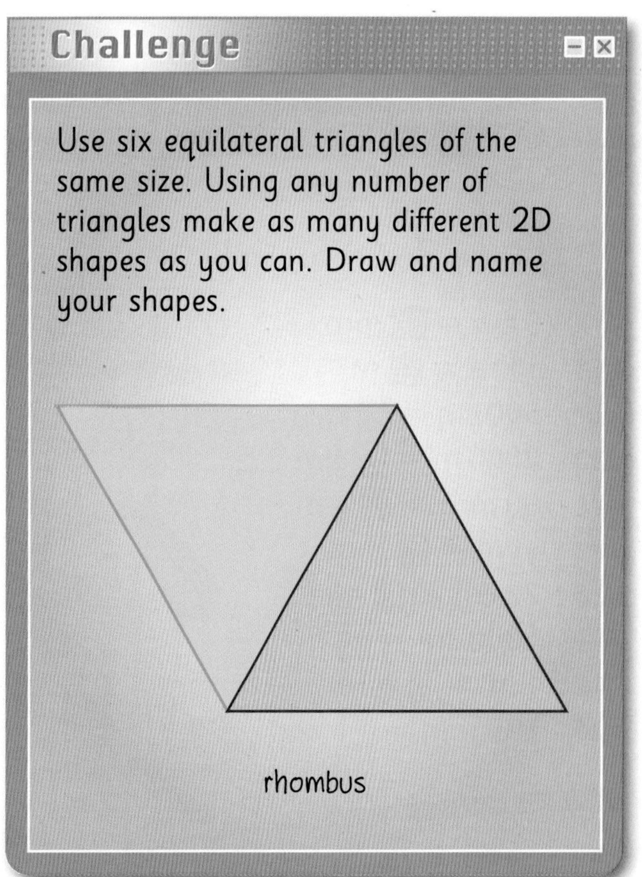

Length

A **Estimate the lengths of these objects to the nearest half cm. Measure each length to the nearest half cm. Find the difference between each estimate and measurement.**

1 estimate 6 cm, measure $5\frac{1}{2}$ cm, difference $\frac{1}{2}$ cm

1 ERASER

2

3

4

5

6 width of a page in this book

7 your hand span

B **Write these distances in metres and centimetres.**

8 1 m 68 cm

8 168 cm
9 142 cm
10 195 cm
11 106 cm
12 201 cm
13 382 cm
14 490 cm
15 301 cm
16 516 cm

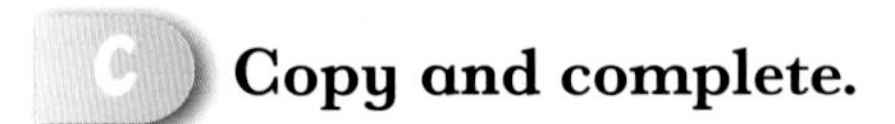

C **Copy and complete.**

17 60 mm

17 6 cm = ☐ mm
18 1 km = ☐ m
19 500 cm = ☐ m
20 2000 m = ☐ km
21 80 mm = ☐ cm
22 $\frac{1}{2}$ cm = ☐ mm
23 22 cm = ☐ mm
24 1000 cm = ☐ m
25 1 m = ☐ mm
26 190 mm = ☐ cm

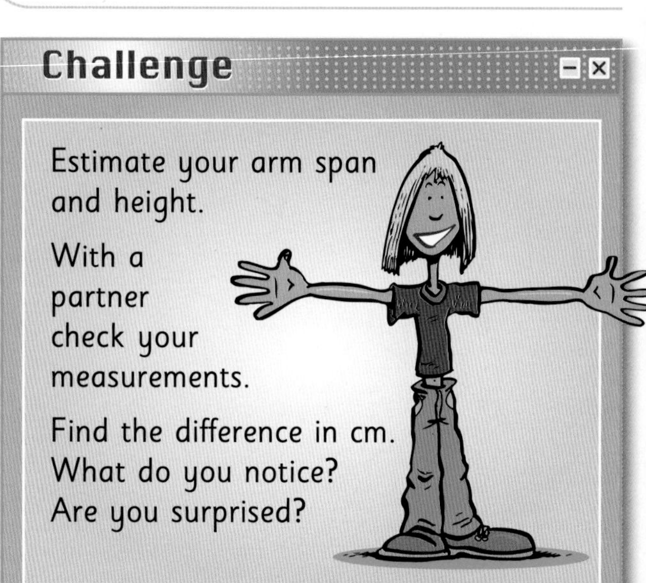

Challenge

Estimate your arm span and height.

With a partner check your measurements.

Find the difference in cm. What do you notice? Are you surprised?

Do you know the relationships between millimetres, centimetres, metres and kilometres?

A Write these heights in metres.

1 1 m 73 cm
2 1 m 57 cm
3 1 m 22 cm
4 153 cm
5 1 m 1 cm
6 97 cm
7 186 cm
8 1m 10 cm
9 1m 36 cm
10 149 cm
11 1m 30 cm

1 1·73 m

B Write these lengths in metres and centimetres.

12 2·85 m
13 1·86 m
14 4·23 m
15 2·30 m
16 2·13 m
17 6·04 m
18 3·01 m
19 16·83 m
20 10·04 m
21 20·63 m
22 18·90 m

C Write these lengths in centimetres.

23 1·29 m
24 1·35 m
25 1·84 m
26 7·37 m
27 4·03 m
28 $\frac{1}{2}$ m
29 0·8 m
30 $\frac{1}{4}$ m
31 $\frac{6}{10}$ m
32 0·75 m
33 $\frac{9}{10}$ m

23 129 cm

D Write which unit of measurement you would use for these lengths.

34 height of your classroom
35 distance from London to Liverpool
36 length of a pen
37 width of a fingernail
38 height of your school
39 width of a page
40 thickness of a worm

34 metres

E Solve these problems.

41 A length of material is 2 m long. How much is left after 1 m 80 cm is cut off?

42 A piece of string is 1 m long. 38 cm is cut off and then another $\frac{1}{4}$ m. What length of string is left?

43 One side of a square is 25 mm. What is the perimeter of the square?

44 Jeff is half as tall as Harry. Harry is 1 m 64 cm tall. How tall is Jeff?

Challenge

a A snake is 1m 65 cm long. If its nose is at one end of your table, will its tail hang over the other end?
b Is your stride longer than the perimeter of this page?
c A robot is 1·9 m high and 0·86 m wide. Will it fit through your classroom door?

Length

A **There are six flies hiding on this page. Estimate, then measure how many millimetres from the centre of the spider's web each fly is? Measure to their noses.**

1. red fly
2. green fly
3. blue fly
4. orange fly
5. yellow fly
6. pink fly

1 red fly 119 mm

B **Solve these word problems.**

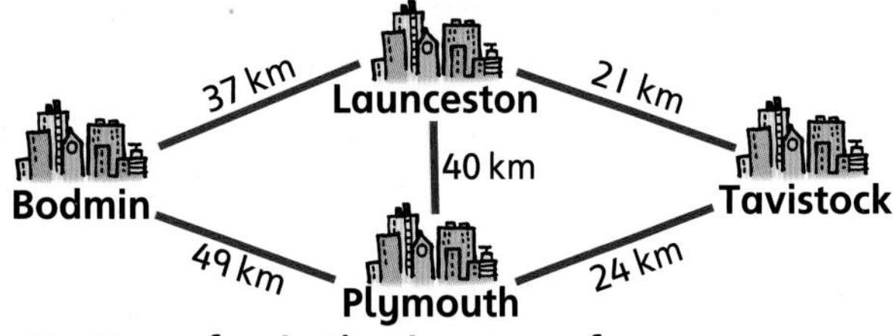

7. How far is the journey from Plymouth to Launceston via Bodmin?
8. How far is the journey from Tavistock to Bodmin via Plymouth?
9. Starting at Bodmin what is the shortest journey to visit all four towns once?
10. A sunflower is 1·46 m tall. A daffodil is 0·57 m tall. How many centimetres taller than the daffodil is the sunflower?
11. Daisy's best high jump is 93 cm. The World Champion can jump 2·23 m. How much higher than Daisy's jump is this?

12. A snail crawled 98·5 cm. How much further must it crawl to travel 1·02 m?
13. If these three blocks are placed end to end, how many centimetres will they stretch?

1·73 m

79 cm

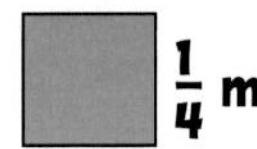

$\frac{1}{4}$ m

14. What is the difference in length between the longest and shortest blocks?
15. Draw a line that is 17 mm shorter than 6 cm.
16. Paving slabs are 46 cm long. Six are placed end to end to make a path. How long is the path?

Challenge

Using appropriate units, measure as accurately as you can:

a the distance five pencils stretch when placed end to end

b the perimeter of your classroom

c the thickness of three different books.

Can you solve problems using km, m, cm and mm?

You need: • a partner • a dice • a ruler • a tape measure • counters

START
Throw again

96 cm

0·82 cm

29 cm

$\frac{1}{4}$ m

0·54 cm

160 mm

$\frac{1}{2}$ m

300 mm

1 m 2 cm

65 cm

82 cm

0·42 cm

0·75 m

26 cm

60 cm

0·63 m

$\frac{3}{4}$ m

58 cm

1 m

200 mm

Each place a counter on **Start**.
Take turns to roll the dice and count round the track.

Estimate the distance shown and place two counters this far apart on your table.

Estimate within	Colour of square
10 cm	blue
5 cm	blue or green
3 cm	blue, green or orange
1 cm	any colour

Measure the distance between them and check how close the estimate was.
Use a counter to cover a matching square on the grid.

The first player with three counters in a straight line is the **winner**.

Perimeter and co-ordinates

A Estimate how far each ant has walked from the anthill to the nearest $\frac{1}{2}$ cm. Measure each distance. Copy and complete the table.

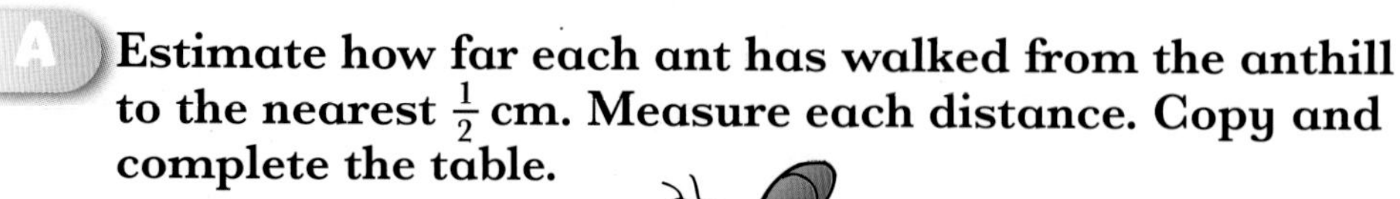

	Ant	Estimate	Measure	Difference
1	Eater			
2	Elope			
3	Enna			
4	Eek			

B Measure, then work out the perimeter of each shape.

5 8 cm

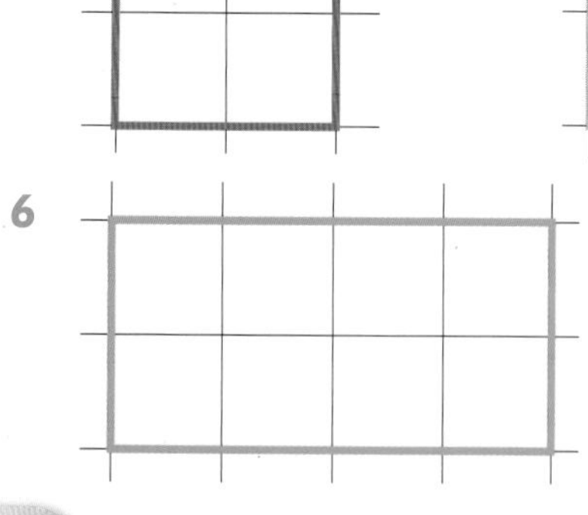
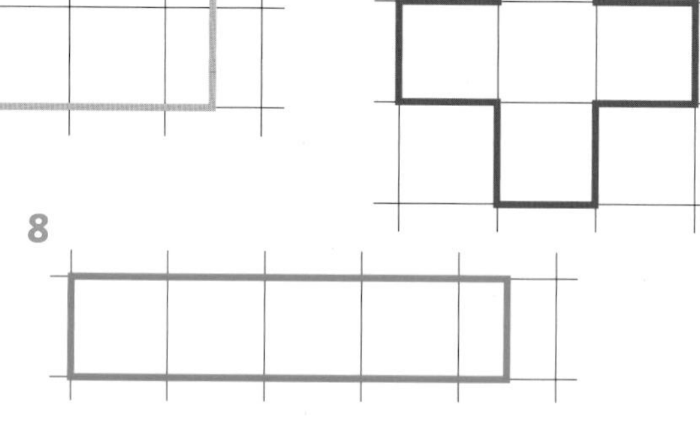

5 7 9 6 8

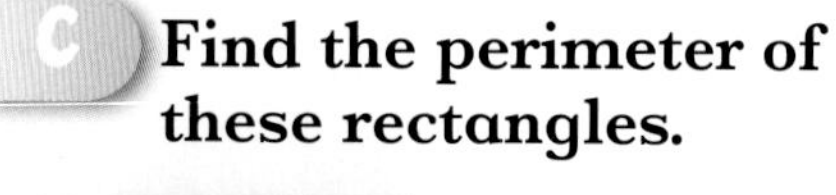
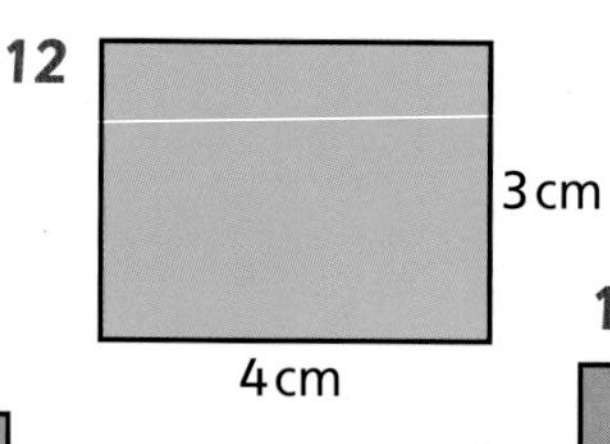

C Find the perimeter of these rectangles.

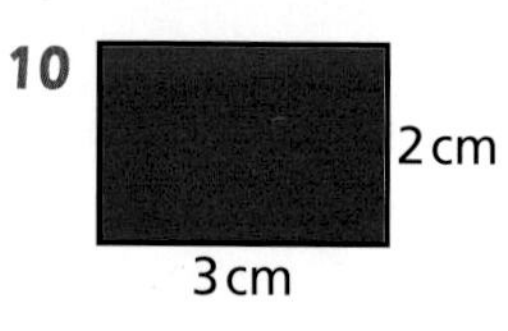

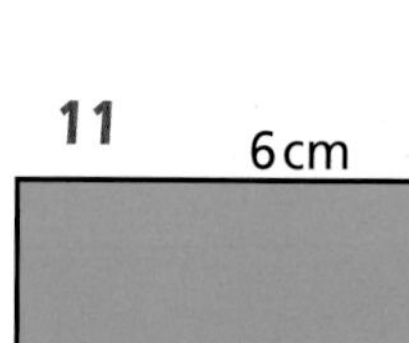

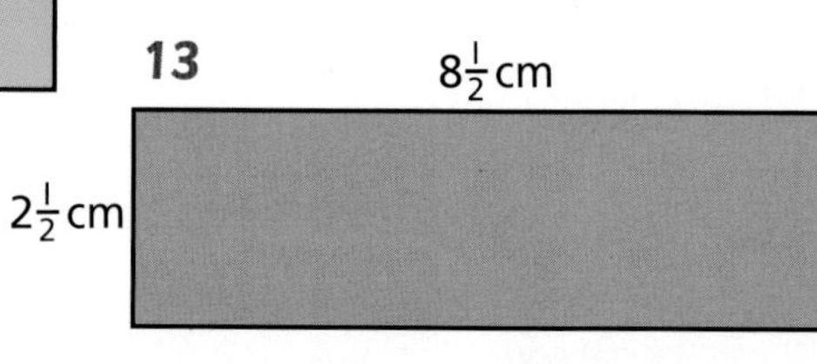

Challenge

On cm-squared paper draw shapes with these perimeters:

a 10 cm **b** 16 cm **c** $15\frac{1}{2}$ cm **d** $5\frac{1}{2}$ cm **e** 95 mm

Can you measure perimeters?

A Estimate, then measure the perimeter of each shape.

1 estimate 10 cm, measure 8 cm

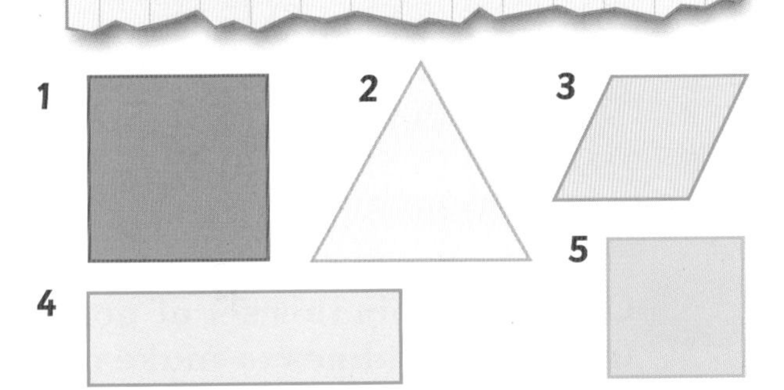

B Write the perimeter of squares with these side lengths.

6 5 cm
7 9 cm
8 12 cm
9 15 mm
10 24 mm
11 $6\frac{1}{2}$ cm
12 120 mm

C Find the length of one side of squares with these perimeters.

13 28 cm
14 40 cm
15 80 cm
16 36 mm
17 48 mm
18 72 cm
19 1 m

D Copy and complete.

	length of rectangle	width of rectangle	perimeter
20	5 cm	6 cm	22 cm
21	7 cm	4 cm	
22	8 cm		24 cm
23		5 cm	27 cm

E Work out the length of each side of squares with these perimeters. Write lengths in cm and mm.

24 4 cm 5 mm

24 18 cm
25 26 cm
26 34 cm
27 58 cm
28 90 cm
29 $\frac{1}{2}$ m

Challenge

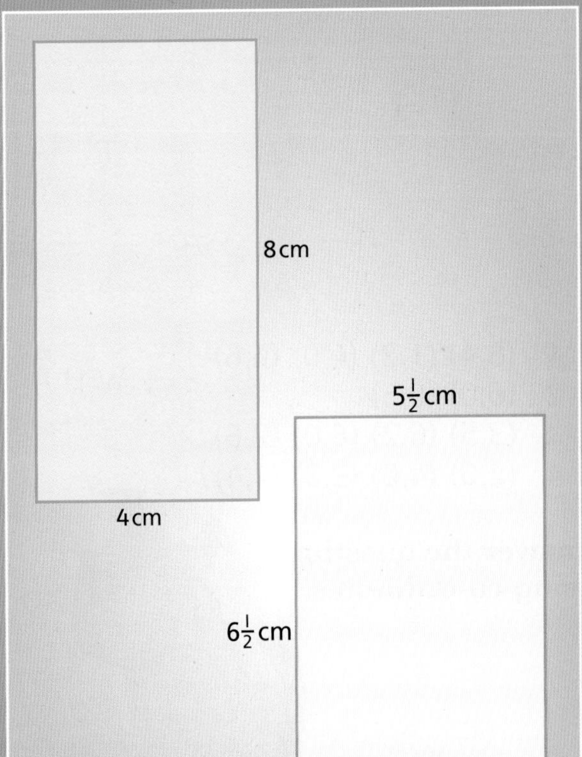

The perimeter of both rectangles is 24 cm. On cm-squared paper, draw some other rectangles with a perimeter of 24 cm. How many can you find?

Can you work out the perimeter of a rectangle from the lengths of two different sides?

Perimeter and co-ordinates

A Find the letter at each position. Work out the secret message.

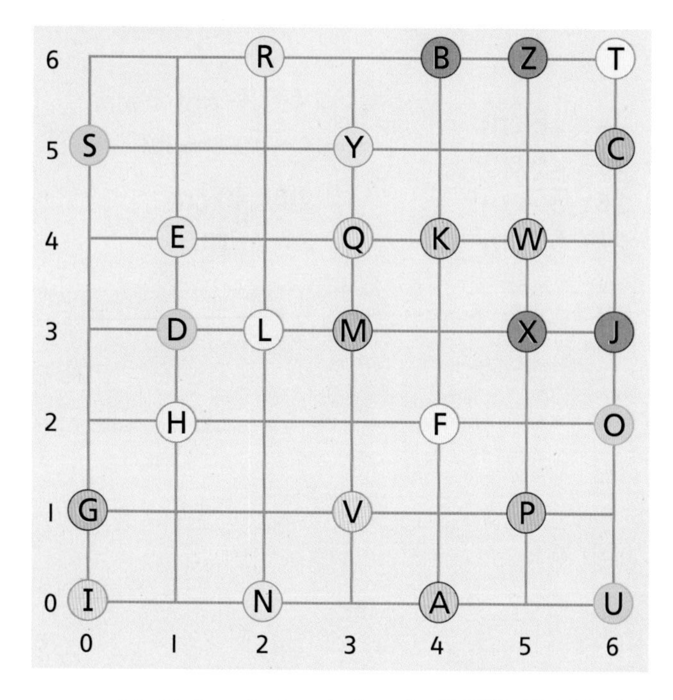

1 (5,4) (1,2) (4,0) (6,6)
2 (0,0) (0,5)
3 (3,5) (6,2) (6,0) (2,6)
4 (2,0) (4,0) (3,3) (1,4)?

1 WHAT

Answer the question using co-ordinates.

B Join each set of points in order. Write the name of the shape you find.

5 rectangle

5 (2,1) (5,1) (5,5) (2,5) (2,1)
6 (3,4) (0,1) (6,1) (3,4)
7 (5,2) (6,3) (5,5) (3,5) (3,3) (5,2)
8 (0,0) (5,0) (5,5) (0,5) (0,0)
9 (2,6) (2,0) (6,0) (2,6)

C On a grid, join this set of points with straight lines to make a picture. What do you see?

10 (0,13) (2,12) (2,11) (0,12) (0,3) (6,3) (6,6) (3,6) (3,3) (4,0) (7,0) (6,3) (9,3) (9,10) (8,10) (8,9) (7,9) (7,10) (6,10) (6,9) (5,9) (5,10) (4,10) (4,9) (3,9) (3,10) (2,10) (2,9) (1,9) (1,10) (0,10)

Challenge

Work out a set of co-ordinates that will make a picture when joined with straight lines. Give a partner your set of co-ordinates to check. Can they draw your picture?

Can you find the position of a point on a grid if you know its co-ordinates?

Review 1

A Copy and complete.

1 32 + 14
2 47 + 18
3 56 + 48
4 64 − 21
5 90 − 16
6 106 + 78
7 327 + 68
8 243 − 87

B Solve these problems.

9 If you have £6·50 and spend £2·85, how much is left?

10 John buys three toys costing £1·50, 85p and £2·75. How much change will he get from £10?

C Match the shapes to their names.

regular pentagon
rhombus
equilateral triangle
irregular hexagon
isosceles triangle

11
12
13
14
15

Write which shapes are symmetrical.

D Write which length is:

16 the longest
17 the shortest
18 $4\frac{1}{2}$ cm
19 between 20 cm and 30 cm
20 3 mm
21 70 cm
22 $\frac{1}{2}$ km.

E Write the perimeter of each shape.

23 square, side 20 mm
24 rectangle, sides 4 m and $3\frac{1}{2}$ m
25 regular heptagon, side 7 cm

F Write the co-ordinates of each coloured dot.

26 black
27 orange
28 red
29 green
30 blue
31 brown

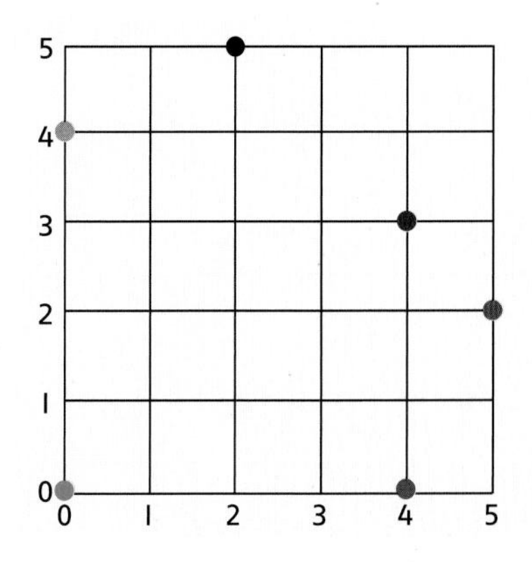

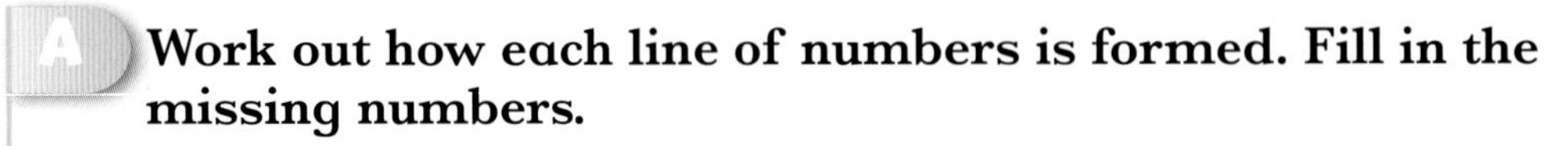

A Work out how each line of numbers is formed. Fill in the missing numbers.

1 21, 24, 27

1. 9, 12, 15, 18, ___, ___, ___
2. 7, 11, 15, 19, ___, ___, ___
3. 16, 21, 26, 31, ___, ___, ___
4. 32, 29, 26, 23, ___, ___, ___
5. 66, 76, 86, 96, ___, ___, ___
6. 428, 438, 448, 458, ___, ___, ___
7. 974, 874, 774, 674, ___, ___, ___
8. 92, 94, 96, 98, ___, ___, ___
9. 341, 331, 321, 311, ___, ___, ___

B Write the rule for each sequence. Fill in the missing numbers.

10 add 6 each time: 41, 47, 53

10. 17, 23, 29, 35, ___, ___, ___
11. 24, 32, 40, 48, ___, ___, ___
12. 92, 87, 82, 77, ___, ___, ___
13. 143, 152, 161, 170, ___, ___, ___
14. 265, 276, 287, 298, ___, ___, ___
15. 841, 821, 801, 781, ___, ___, ___
16. 1040, 1030, 1020, 1010, ___, ___, ___

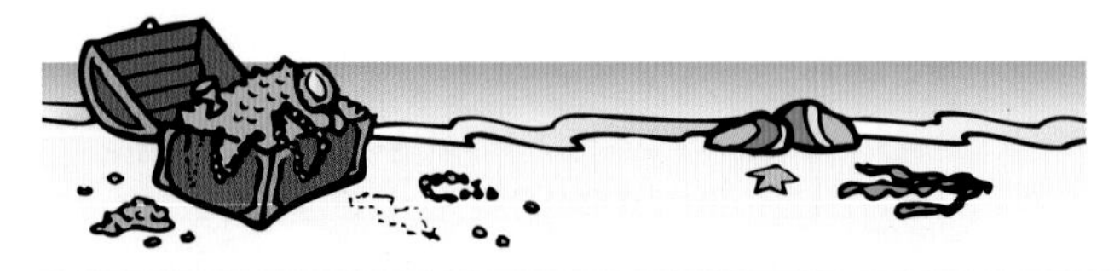

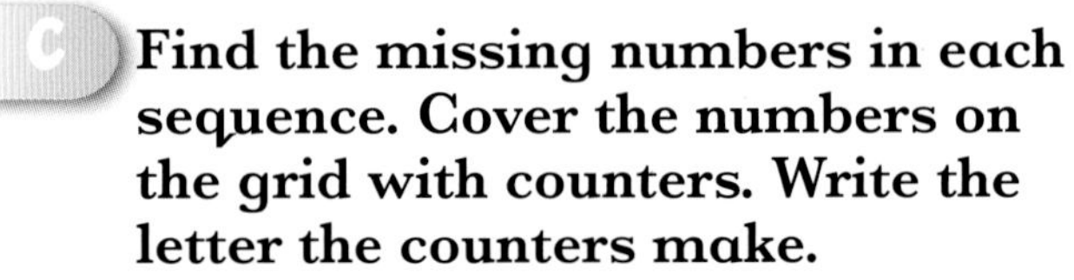

C Find the missing numbers in each sequence. Cover the numbers on the grid with counters. Write the letter the counters make.

38	29	−3	8	0
19	−1	48	15	80
104	62	42	1	72
7	46	16	−2	56

17. 26, 32, ___, 44, 50, ___
18. ___, 35, 28, 21, 14, ___, ___
19. ___, 96, 88, ___, ___, 64, 56
20. ___, ___, 11, 7, 3, ___, −5

Challenge

Will the number 100 ever appear in each of these sequences? Explain your answers.

a 42, 44, 46, 48, 50, 52, ___, ___

b 425, 400, 375, 350, 325, ___, ___

c 48, 52, 56, 60, 64, ___, ___

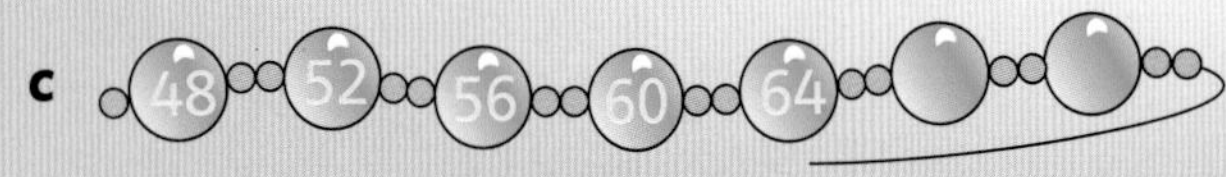

d −40, −30, −20, −10, 0, ___, ___

e 1000, 998, 996, 994, 992, ___, ___

Can you find the rule and extend a number sequence?

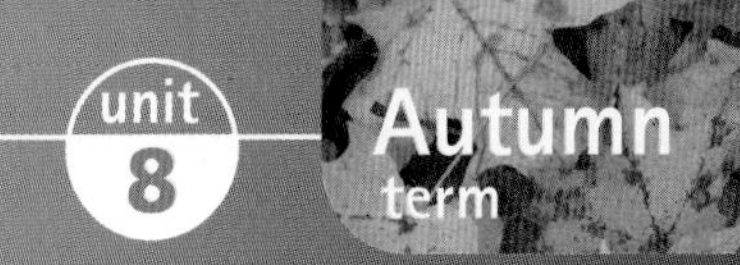

A Copy and complete each sequence.

1 $^-$12, $^-$19

1 16, 9, 2, $^-$5, __, __
2 25, 19, 13, 7, __, __, __
3 50, 40, __, 20, __, __, __
4 17, 9, __, $^-$7, __, __
5 25, 16, __, __, $^-$11, $^-$20, __
6 __, __, $^-$6, $^-$13, $^-$20, __
7 __, 0, __, $^-$14, $^-$21, $^-$28

B Make up a sequence of numbers that ends with each number.

8 32, 28, 24, 20, 16, 12

8 12
9 18
10 0
11 101
12 1000
13 $^-$3
14 $^-$15

C Write odd or even number for each.

15 even number

15 sum of 2 even numbers
16 sum of 2 odd numbers
17 sum of odd number and even number
18 sum of 3 even numbers
19 sum of 3 odd numbers
20 double an odd number
21 double an even number

D Write odd or even.

22 the next number after 927
23 the number immediately before 801
24 the number 5 more than 666
25 the number 8 less than 1000
26 the number 9961
27 the sum of 999 and any other odd number

Challenge

Throw four dice together.
If the numbers are all odd or all even, throw again. Use the numbers you throw to make:

a the largest odd number
b the smallest even number
c the closest number to 3000
d the number nearest to 5000.

A **This grid is called a magic square. Answer these questions about it.**

8	3	4
1	5	9
6	7	2

1 What is the total of each set of three numbers in the direction of each arrow?

2 Why do you think the square is called a magic square?

3 What missing numbers complete this magic square so that each line totals 24?

		10
	8	3
		11

B **Solve these problems.**

4 0 + 1 + 2 = 3
1 + 2 + 3 = 6

4 Add together each set of three consecutive numbers from 0 to 20.

5 Write about any patterns you find.

C **Find three consecutive numbers that add to make each total.**

6 72 7 51 8 87 9 99

D **Look at this number pyramid. Solve these problems.**

10 top row: 1
2nd row: 1 + 1 = 2

10 Work out the total of the numbers in each complete row of the pyramid.

					1					
				1		1				
			1		2		1			
		1		3		3		1		
	1		4		6		4		1	
		10				5				

11 Add together next-door numbers in the 5th row to work out the numbers for the 6th row.

12 Find the total of the numbers in the 6th row.

13 Write about any patterns you find.

Challenge

A spy is trying to crack a code. Each of the numbers 1, 2, 3, 4 and 5 has been replaced by a symbol. Work out which symbol replaces each number.

Clues:

✪ + ✪ = ■ ■ + ◆ = ✷
✷ – ▲ = ✪ ■ + ■ = ▲
▲ – ◆ = ✪ ■ × ■ = ▲

Answer these calculations in numbers.

a ✪■✷ + ◆✪▲
b ✷▲✪ – ■◆✷

Can you solve puzzles and find patterns?

A Copy and complete.

1 4 × 5
2 6 × 3
3 3 × 5
4 20 ÷ 2
5 $\frac{1}{2}$ of 16
6 5 × 10
7 $\frac{1}{2}$ of 60
8 5 × 5
9 double 25
10 60 ÷ 10
11 32 ÷ 4

B Solve these problems.

12 A piece of string is 18 m long. If it is cut into three equal pieces, how long is each piece?

13 Crumpets are packed in fives. How many packs can be made with 35 crumpets?

14 Anish spent half of his £1·60 pocket money. How much did he spend?

C Double the numbers on the red rockets. Halve the numbers on the blue rockets.

15 14
16 16
17 24
18 25
19 34
20 84
21 126
22 68

D Complete these.

23 16 × 20
24 14 × 10
25 15 × 4
26 22 × 4
27 14 × 8
28 12 × 25
29 21 × 20
30 14 × 20
31 13 × 30

Challenge

Work out the clues and follow the trail to find who stole the jewels.

13 × 4: 52, 42
16 × 10: 160, 1600
21 × 4: 48, 84
13 × 4: 140, 40
23 × 20: 460, 232
11 × 25: 275, 250
14 × 30: 420, 340
32 × 4: 128, 92
9 × 40: 36, 360
26 × 20: 260, 520
16 × 30: 48, 480
18 × 8: 164, 144
12 × 60: 720, 620

Basher Pincher Sneaky Dozy Lofty

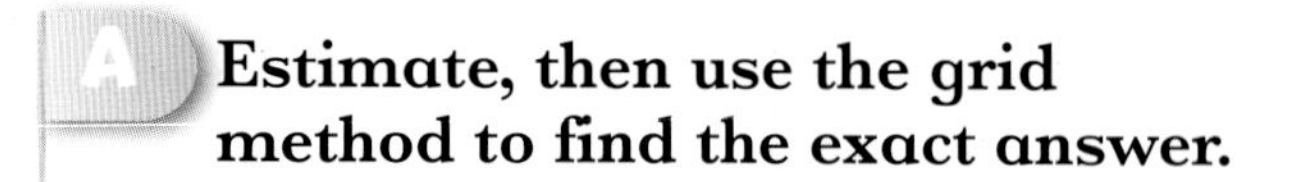

A Estimate, then use the grid method to find the exact answer.

1 42 × 3
2 29 × 2
3 36 × 3
4 45 × 4
5 62 × 5
6 53 × 4
7 27 × 5
8 67 × 3
9 78 × 4

1 estimate = 120

×	40	2	
3	120	6	126

B Write each amount.

10 $\frac{1}{2}$ of £46
11 $\frac{1}{2}$ of £36
12 $\frac{1}{2}$ of £52
13 $\frac{1}{2}$ of £220
14 $\frac{1}{4}$ of £32
15 $\frac{1}{4}$ of £400
16 $\frac{1}{4}$ of 68p

C Write how many each and how many left over when shared equally.

17 8 cows each and 2 left over

17 26 cows
3 farmers

19 111 flies
4 spiders

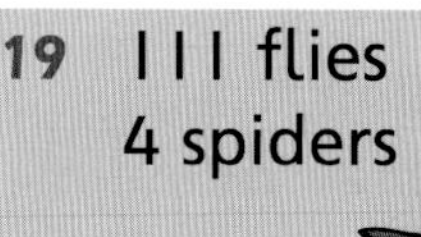

18 56 comics
5 classes

20 43 balloons
3 clowns

Challenge

16	×	3	=	
÷		×		÷
4	×		=	4
=		=		=
4	×		=	

	÷	4	=	15
÷		×		÷
3	×		=	3
=		=		=
	÷		=	

Investigate how these grids work. Copy and complete each grid. Write numbers in each empty space.

Try to construct your own grid.

Can you use different methods to solve division problems?

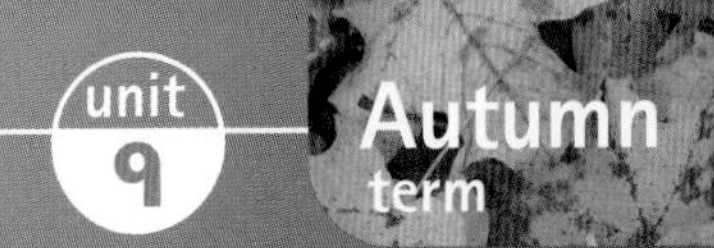

A Find who catches each item.

1 6 + 6 + 6 + 6 + 6 = 6 × 5
Jason catches the treasure chest

1 Jason 6 + 6 + 6 + 6 + 6
2 Jilly product of 9 and 6
3 Jack 12 × 4
4 Joe 26 × 12
5 Jenny product of 7 and 5
6 Jean double 89
7 Jonah 9 × 5
8 Jonty half of 354

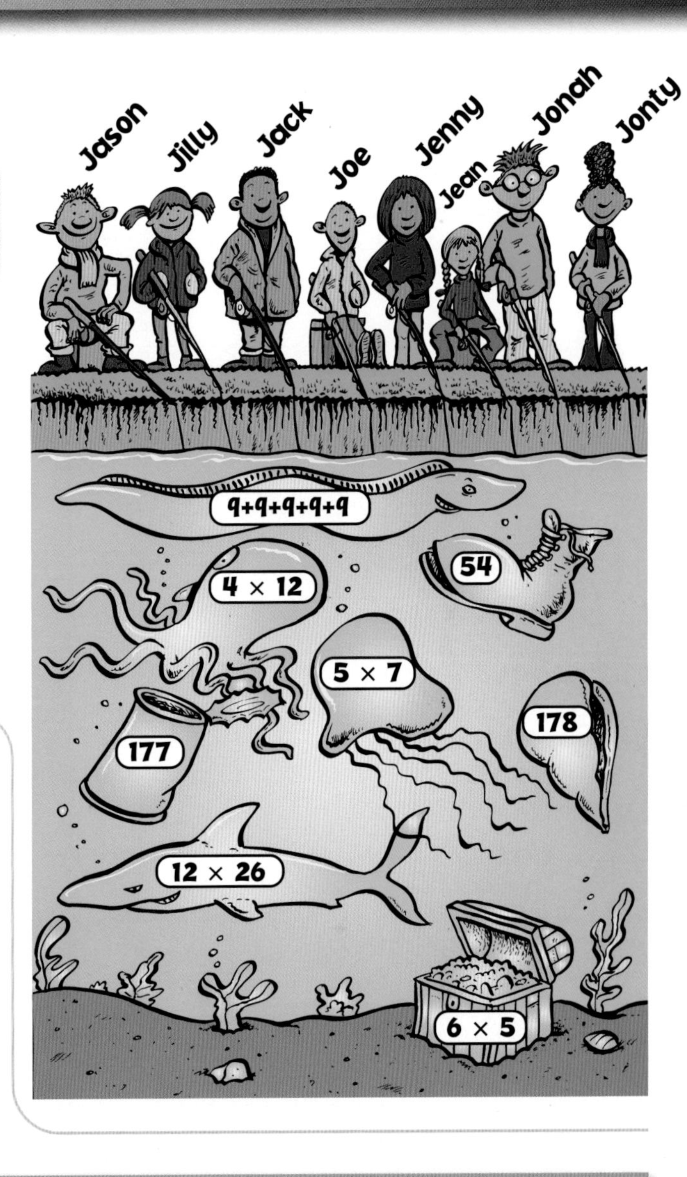

B Complete these.

9 10 r 1

9 31 ÷ 3
10 50 ÷ 5
11 46 ÷ 4
12 52 ÷ 3
13 61 ÷ 7
14 53 ÷ 9
15 75 ÷ 8
16 46 ÷ 4

Challenge

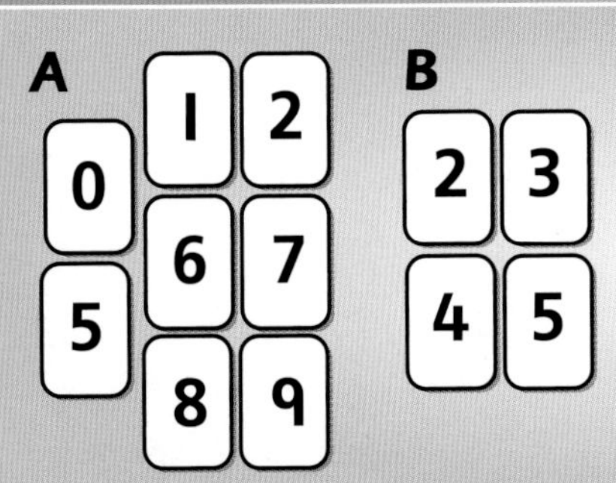

Use the cards shown. Shuffle each set and place them face down. Take two cards from set A to make a 2-digit number. Take a card from set B. Multiply the numbers together □ □ × □. Score the digit total of the answer. E.g. 51 × 2 = 102, 1 + 0 + 2 = 3 → score 3.
Take four turns. What is your total score? Try again. What is your highest score?

Perimeter points

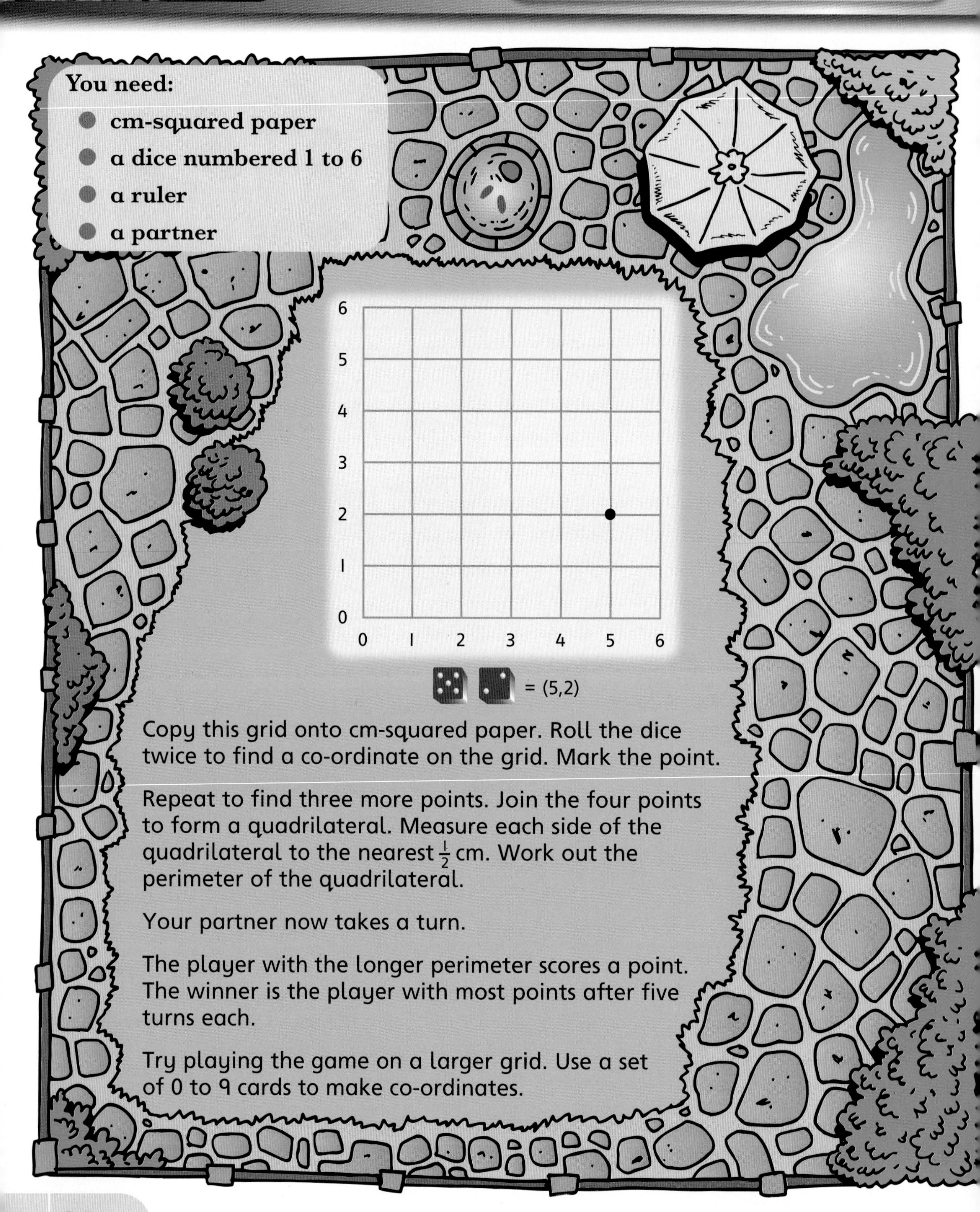

You need:

- cm-squared paper
- a dice numbered 1 to 6
- a ruler
- a partner

Copy this grid onto cm-squared paper. Roll the dice twice to find a co-ordinate on the grid. Mark the point.

Repeat to find three more points. Join the four points to form a quadrilateral. Measure each side of the quadrilateral to the nearest $\frac{1}{2}$ cm. Work out the perimeter of the quadrilateral.

Your partner now takes a turn.

The player with the longer perimeter scores a point. The winner is the player with most points after five turns each.

Try playing the game on a larger grid. Use a set of 0 to 9 cards to make co-ordinates.

A Write how much you spend if you buy these items.

1 4 badges
2 2 T-shirts
3 3 caps
4 1 T-shirt and 2 badges
5 1 scarf and 1 cap
6 1 ball and 1 scarf
7 2 scarves and 1 T-shirt
8 10 badges and 1 cap
9 1 of each item
10 4 T-shirts
11 4 balls

B Solve these problems.

12 Rangers fans wear 4 badges on their caps. How many badges are needed for 19 caps?

13 T-shirts are packed in boxes of 8. How many T-shirts in 13 boxes?

14 Badges are packed in boxes of 25. What is the cost of a box of badges?

15 Sasha is given £2·50 pocket money each week. How many weeks will it take her to save up for a scarf and a ball?

16 Leo has saved £50. He buys 9 badges and 3 caps. How much is left?

17 How many fans can have a cap with 4 badges if there are 72 badges?

Challenge

Mr Patel went shopping on the High Street. The diagram shows how much he spent in each shop and what amount he paid with.

shop	a	b	c	d	e
spent	£2·65	£15·62	96p	£5·55	£2·72
paid with	£5	£20	£5	£6	£5

Look at Mr Patel's notes about how much change he received to work out how much he spent in each shop.

shop	change
Toy shop	£4·38
Bakery	£2·35
Sweet shop	£2·28
Chemist	£0·45
Newsagent	£4·04

A Complete these.

1 38×10
2 24×100
3 63×10
4 13×100
5 59×100
6 61×10
7 40×100

B The answers to these problems are hidden in the letters below. Write letters to match each answer to find a secret word.

8 Mrs Briggs buys 32 apples. Her family eats a quarter of them. How many are left?

9 Mr Button buys 2 screwdrivers for £5·40 each and 10 packs of screws at 22p a pack. How many pounds change does he get from £20?

10 Alice thinks of a number and then adds 26. Her answer is 58. What was her number?

11 There are 82 sweets in a jar. How many are left after 9 people take 5 sweets each?

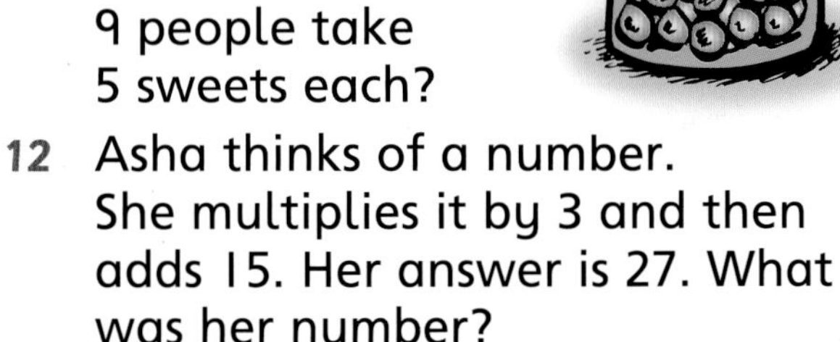

12 Asha thinks of a number. She multiplies it by 3 and then adds 15. Her answer is 27. What was her number?

13 Jess has 48 marbles. Half are red and one third are blue. How many marbles are not red or blue?

A 4, B 16, D 9, E 37, G 18, H 20, K 8, L 30, M 14
N 36, O 6, P 2, Q 7, S 24, U 32, V 42, W 10

Challenge

Here are four answers. Make up a **really difficult** problem for each answer. Ask a partner to try to solve your problems.

a 35
b £1·60
c 27
d £1·68

A **Solve these problems. Show your working clearly. Explain how to check each answer by doing a different calculation.**

1 The cost of 4 oranges is 88p. What do 14 oranges cost?

2 Flowers cost 34p each. How many flowers can Daisy buy with a £5 note?

3 4000 people enter a competition to become a pop star. After three days 3643 people have been told they are not good enough. How many are left in the competition?

4 Full packets hold 16 mints. Mandy buys 4 packets and eats 9 mints from 1 pack. How many mints are left?

1 1 orange costs 88p ÷ 4 = 22p
10 oranges cost 10 × 22p = £2·20
Cost of 10 oranges + cost of 4 oranges = cost of 14 oranges
£2·20 + 88p = £3·08

Check:
4 oranges cost 88p
8 oranges cost £1·76
12 oranges cost £1·76 + 88p = £2·64
2 oranges cost $\frac{1}{2}$ of 88p = 44p
14 oranges cost £2·64 + 44p = £3·08

5 Zoe wants to buy a computer game for £7·65, a netball for £6·24 and a radio for £23·64. She has saved £19·55. How much more money does she need?

6 Red balloons cost 24p, green balloons cost 22p, and yellow balloons cost 20p. Sally buys 30 balloons. Half are red, one third are yellow and the rest are green. How much does Sally spend?

Challenge

An open-top bus ride in London costs £17 per adult and half price for a child. A family ticket for 2 adults and 2 children costs £50. 1 adult and 1 child can travel for £24. What is the cheapest fare for 3 adults and 2 children?

Can you check with an equivalent calculation?

A What fraction of each shape is shaded?

1 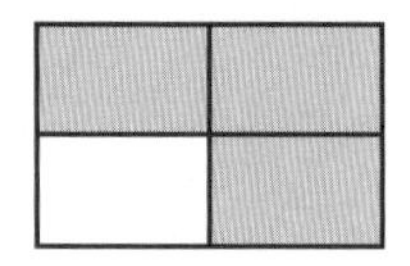

1 $\frac{3}{4}$

2 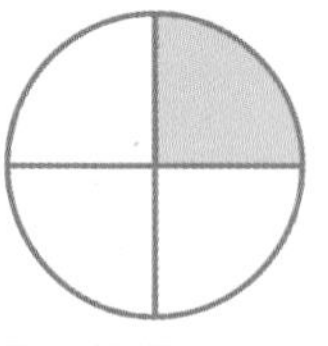

5 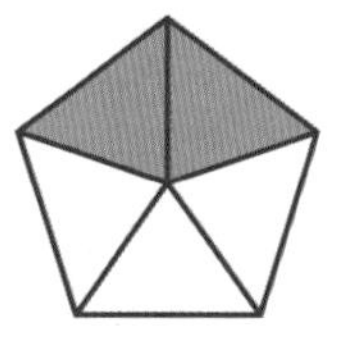

3 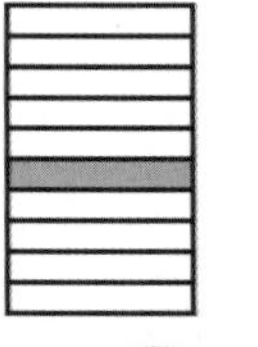

6 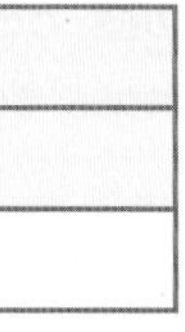

4 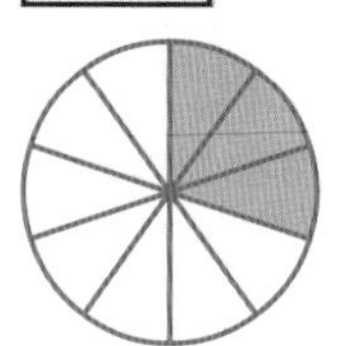

7 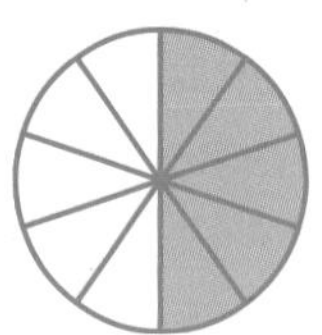

B What fraction of each group of ants is red?

8  

8 $\frac{5}{8}$ are red

9, 11 

10, 12 

C Write the number covered by each yellow counter.

13 $\frac{2}{10} = \frac{1}{5}$

13 $\frac{2}{10} = \frac{●}{5}$ 15 $\frac{2}{3} = \frac{●}{6}$ 17 $\frac{6}{10} = \frac{●}{5}$

14 $\frac{4}{5} = \frac{●}{10}$ 16 $\frac{4}{8} = \frac{1}{●}$ 18 $\frac{4}{4} = \frac{2}{●}$

D Copy and complete.

19 4 quarters

19 1 whole = ⬢ quarters
20 1 whole = ✸ sixths
21 1 whole = ◆ fifths
22 1 whole = ● eighths
23 1 whole = ■ halves

Challenge

Work out the fractions missing from each number line.

a 0, $\frac{1}{4}$, 1

b 0, $\frac{3}{5}$, 1

c 0, $\frac{1}{2}$, 1

d 0, 1, $1\frac{1}{3}$, 2

Can you work out equivalent fractions?

A Write which colour dot shows the position of each fraction.

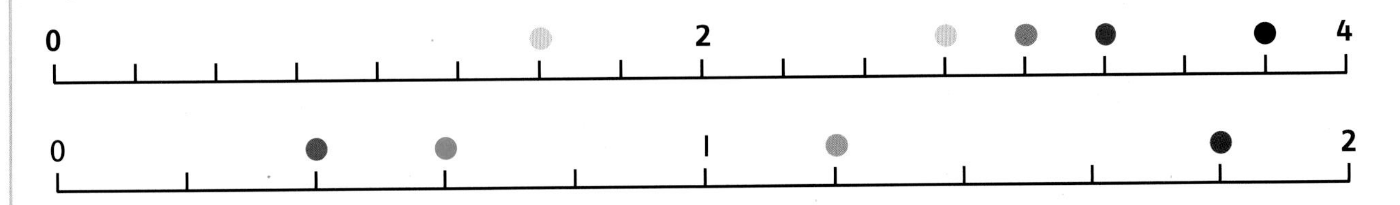

1 $3\frac{1}{4}$

2 $\frac{3}{5}$

3 $1\frac{1}{5}$

4 $1\frac{1}{2}$

5 $2\frac{3}{4}$

6 $1\frac{4}{5}$

7 $3\frac{3}{4}$

1 red

B Write the colour of the section of the number line where each fraction is found.

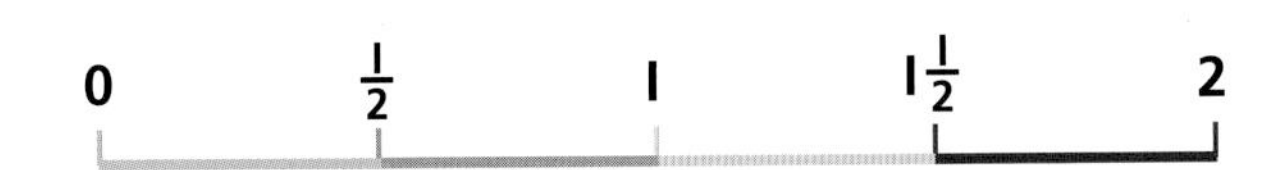

8 $1\frac{1}{3}$

9 $\frac{3}{4}$

10 $1\frac{1}{5}$

11 $\frac{4}{5}$

12 $1\frac{5}{8}$

13 $\frac{7}{10}$

14 $1\frac{3}{5}$

8 yellow

C Write the fraction of each cake that has been eaten.

15 $\frac{4}{5}$ left

16 $\frac{2}{3}$ left

17 $\frac{2}{5}$ left

18 $\frac{1}{10}$ left

19 $\frac{5}{8}$ left

20 $\frac{5}{6}$ left

15 $\frac{1}{5}$ eaten

D There are 10 sweets in a pack. Write how many sweets in each fraction of a pack.

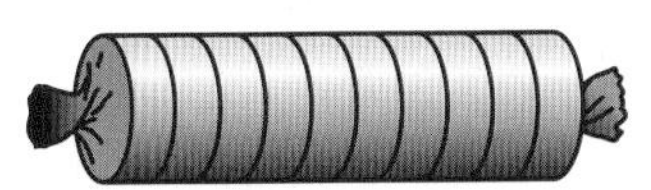

21 $\frac{3}{10}$ of a pack

22 $\frac{7}{10}$ of a pack

23 $\frac{1}{5}$ of a pack

24 $\frac{1}{2}$ a pack

25 $\frac{4}{5}$ of a pack

Challenge

Use cm-squared paper to draw these:

a a rectangle with 15 squares, $\frac{1}{3}$ shaded

b a rectangle with 16 squares, $\frac{3}{4}$ shaded

c a square with 25 squares, $\frac{4}{5}$ shaded

d any shape with 24 squares, $\frac{3}{8}$ shaded.

Can you order fractions and mixed numbers on a number line?

A Write how many tomatoes.

1. $\frac{1}{4}$ of 12 tomatoes
2. $\frac{1}{5}$ of 20 tomatoes
3. $\frac{1}{3}$ of 18 tomatoes
4. $\frac{1}{6}$ of 18 tomatoes
5. $\frac{1}{10}$ of 40 tomatoes

B There are 30 days in June. Find how many days in these.

6. $\frac{2}{3}$ of June
7. $\frac{2}{5}$ of June
8. $\frac{3}{10}$ of June
9. $\frac{3}{5}$ of June
10. $\frac{9}{10}$ of June
11. $\frac{5}{6}$ of June

C Solve these problems. Show your working.

12. There are 20 cars in a garage. $\frac{3}{5}$ are red. How many are not red?
13. Which would you prefer, $\frac{3}{4}$ of £24 or $\frac{2}{3}$ of £30?
14. There are 200 children in school. $\frac{1}{5}$ are in Year 4. How many children are not in Year 4?
15. A blue piece of string is $\frac{5}{8}$ of 40 cm long. A red piece is $\frac{2}{3}$ of 36 cm long. Which is longer?
16. What fraction of the large weight is the smaller weight?

17. Which is the largest amount: $\frac{3}{10}$ of £110, $\frac{14}{100}$ of £200, $\frac{4}{5}$ of £40?

Challenge

Find circles that have the same value as the fractions and mixed numbers below. Work out the hidden message.

$1\frac{2}{8}, 1\frac{6}{8}, \frac{5}{10}, \frac{2}{10}$

$1\frac{25}{100}, 1\frac{1}{2}, \frac{8}{10}, \frac{5}{5}, \frac{2}{6}, \frac{3}{3}, \frac{2}{5}, 1\frac{3}{6}, \frac{4}{8}$

$\frac{2}{10}, \frac{9}{10}, \frac{2}{4}, \frac{20}{100}, 1\frac{5}{10}$

$\frac{3}{5}, 1\frac{175}{100}, \frac{6}{8}, \frac{75}{100}, \frac{8}{8}, \frac{1}{4}, \frac{4}{6}, 1\frac{2}{4}$

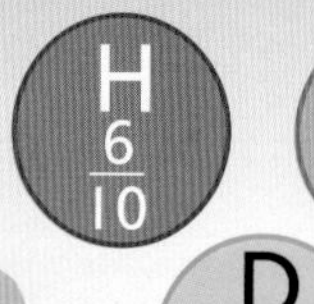

H $\frac{6}{10}$, S $\frac{1}{2}$, L $\frac{2}{3}$, A $\frac{90}{100}$, T $\frac{1}{5}$, D $\frac{4}{5}$, E $1\frac{4}{8}$, R $\frac{3}{4}$, I 1, C $\frac{1}{3}$, B $\frac{2}{8}$, M $1\frac{1}{4}$, O $1\frac{3}{4}$, N $\frac{4}{10}$

Can you work out fractions of numbers and quantities?

You need:

- a calculator

Solve these problems.

Add all your answers together on a calculator.

Turn the calculator upside down to read what the children take home from the beach.

1. $\frac{1}{8}$ of 40
2. $\frac{3}{10}$ of 50
3. Zena thinks of a number. She adds 5 and multiplies her answer by 4 to get 56. What was her number?
4. 26×100
5. Golf balls are packed in tubes of 8. How many golf balls in 13 tubes?
6. $\frac{3}{4}$ of 52
7. Mrs Lucky wins £6000 in a competition. She spends $\frac{2}{3}$ of her money. How many pounds does she have left?

8. There are 25 nails in a pack. How many nails in 25 packs?

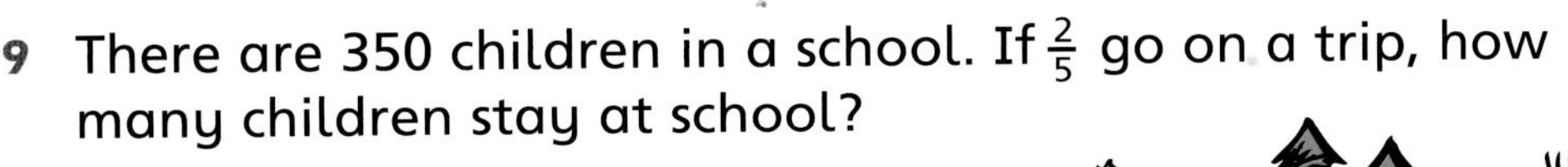

9. There are 350 children in a school. If $\frac{2}{5}$ go on a trip, how many children stay at school?

Write three other facts to link with each of these.

1 9 + 16 = 25, 25 − 9 = 16, 25 − 16 = 9

1 16 + 9 = 25
2 18 + 9 = 27
3 16 + 15 = 31
4 15 + 17 = 32
5 28 − 9 = 19
6 43 − 17 = 26
7 34 + 18 = 52

Write each time.

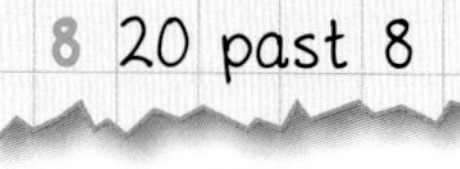

8

9

10

11

12

13

Write three other facts to link with each of these.

14 28 + 65 = 93
15 68 + 39 = 107
16 94 − 38 = 56
17 73 + 81 = 154
18 127 − 88 = 39
19 46 + 64 = 110

Solve these problems.

20 I am thinking of a number. If I add 23 to my number, the answer is 42. What is my number?

21 19 more than my number is 46. What is my number?

22 32 less than my number is 95. What is my number?

23 Simon's pulse rate increases by 38. If it is now 101, what was it before?

24 What is the difference between 95 and 103?

25 The number of children in a school is decreased by 44. There are now 68 children. How many were there before?

Challenge

Use numbers in the yellow box to make additions and subtractions that give answers in the red box. How many answers can you use?

66 − 17 = 49

51 66 48 26 64 17

Answers

31 3 34 117 25 38 49 65 40 43 114

Do you understand that addition is the inverse of subtraction?

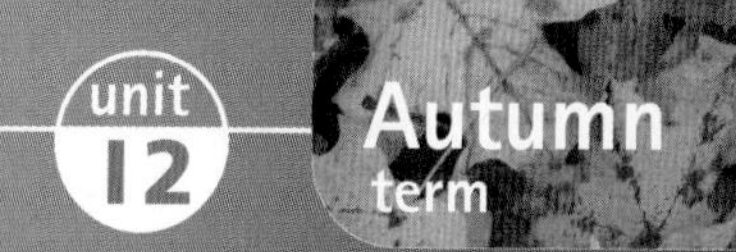

A Answer these.

1 74 − 66
2 62 − 58
3 101 − 97
4 116 − 109
5 427 − 419
6 802 − 793
7 2002 − 1998

B Use a written method to solve these subtractions.

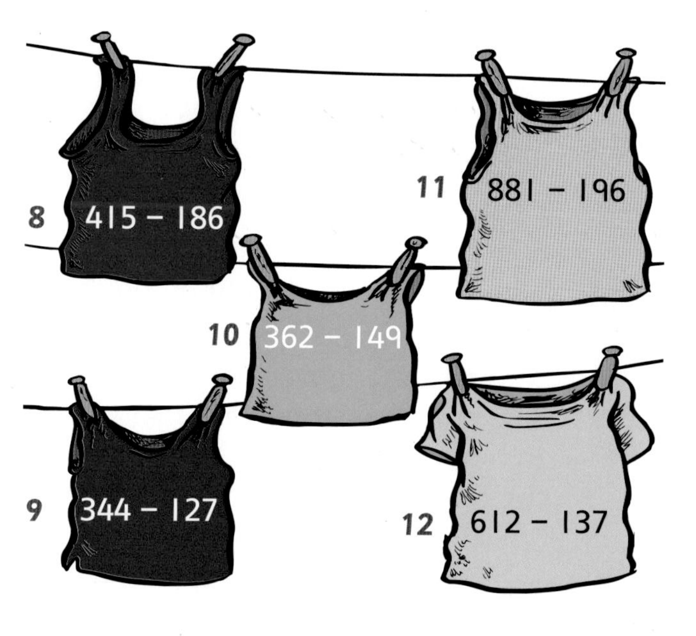

C Work out the difference in scores between each pair of athletes.

DECATHLON RESULTS	
J Silva	7021
S Kinouti	7004
R Day	6995
I Kudwin	6629
U Kanluse	6417
E Isgud	5948

13 Kinouti and Silva
14 Day and Kinouti
15 Kanluse and Kudwin

16 Isgud and Silva
17 Kudwin and Kinouti
18 Kudwin and Isgud

Challenge

Follow different routes through the rabbit warren. Which routes lead to a correct answer?

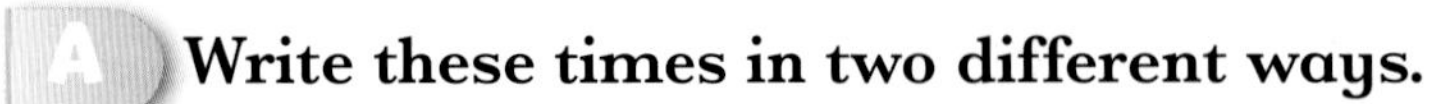

A Write these times in two different ways.

1 

2 

3 

4 

5 

1 23 minutes to 3
2:37

B Use the bus timetable to answer these questions.

BUS TIMETABLE

	Bus 1	Bus 2
Mirton	8:47 a.m.	11:15 a.m.
Batby	8:59 a.m.	11:26 a.m.
Cleckton	9:14 a.m.	11:40 a.m.
Dewsby	9:42 a.m.	12:06 p.m.
Pudfield	10:02 a.m.	12:31 p.m.

6 How long does Bus 1 take to travel from Batby to Dewsby?

7 How long does Bus 2 take to travel from Mirton to Dewsby?

8 How long does Bus 1 take to travel from Cleckton to Pudfield?

9 If you need to be in Dewsby at noon, which bus must you catch from Mirton?

10 Which bus takes longer to travel from Mirton to Pudfield?

C Solve these time problems.

11 It takes Mrs Djemba 24 minutes to walk to the station. If she arrives at 8:32 a.m., what time did she set out?

12 Khaled started swimming at 6:26 p.m. and swam for 40 minutes. What time did he finish?

13 Megan takes 38 minutes to cycle to work. She leaves home at 7:37 a.m. She stops for 8 minutes to buy a paper. What time does she arrive at work?

D Write these 24-hour clock times as a.m. or p.m. times.

14	13:26	18	14:01		
15	15:09	19	21:17		
16	16:52	20	23:58		
17	20:37	21	01:10	22	00:32

14 1:26 p.m.

Challenge

Draw a timeline with clock faces to show what happens during one day in your life.

Can you find times earlier or later than a given time?

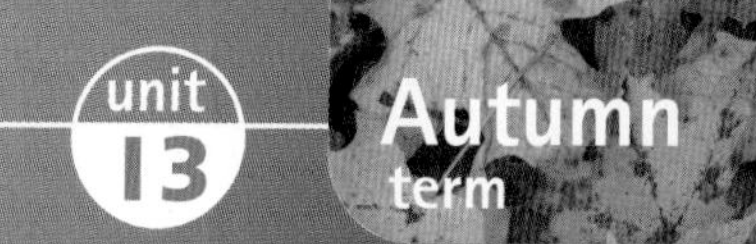

Bird Spotting

A Write the length of each bird.

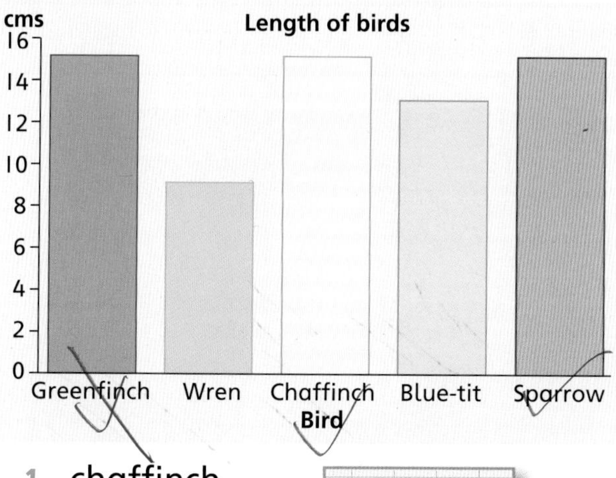

1 chaffinch
2 wren
3 greenfinch
4 blue-tit
5 sparrow
6 Which birds are the same length as the chaffinch?
7 How much longer is a greenfinch than a wren?

1 15 cm

B Write how many of each bird was seen on the bird table.

Friday 12 February: birds seen on bird table between 10:30 a.m. and 11:00 a.m.

Sparrow	𝍸 𝍸 II
Wren	II
Chaffinch	II
Blue-tit	𝍸 𝍸 𝍸 II
Greenfinch	

8 sparrows
9 blue-tits
10 greenfinches
11 wrens
12 chaffinches

8 12

C Write how many of each bird were seen in the garden.

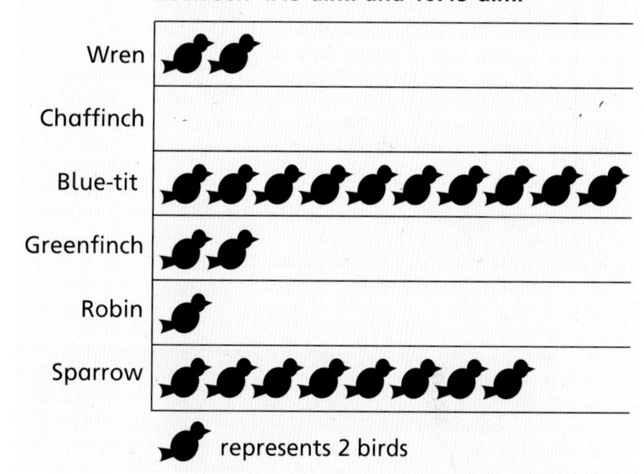

13 blue-tits
14 greenfinches
15 chaffinches
16 sparrows
17 wrens
18 robins

13 20

Challenge

John spotted a bird on the bird table in the garden at 10:40 a.m. It was more than 14 cm in length. Use the information above to work out which bird it was. Write some clues like this to try on a partner.

Can you gather information from tally charts, pictograms and bar charts?

A **Copy and complete this frequency table and bar chart for the animals at Dewsley Zoo.**

Animal	Cheetahs	Rabbits	Sheep	Tigers	Snakes
Number	卌 卌				

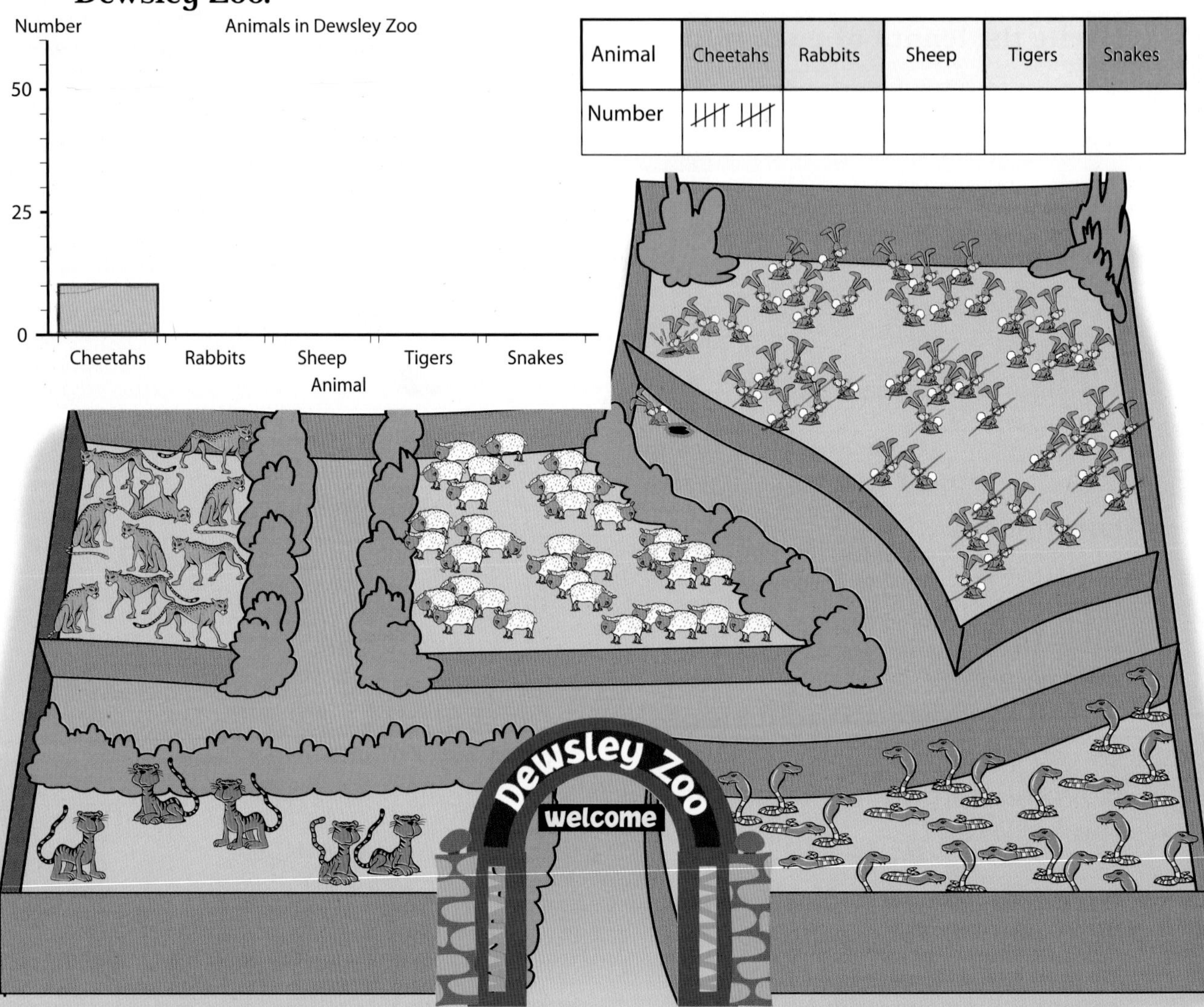

Challenge

This bar chart shows some information, but the title and labels are missing. Suggest three sets of suitable titles and labels for the bar chart.

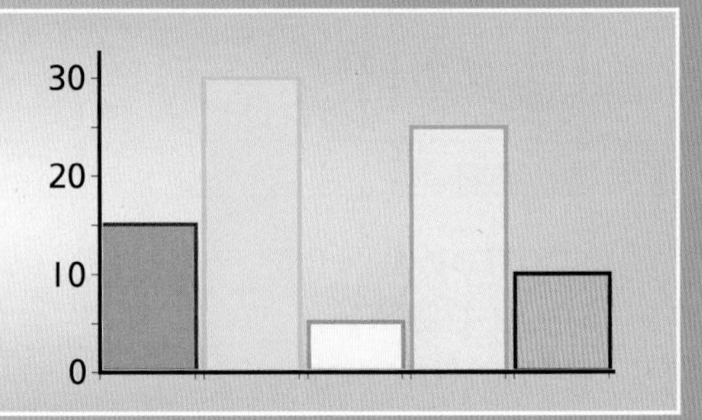

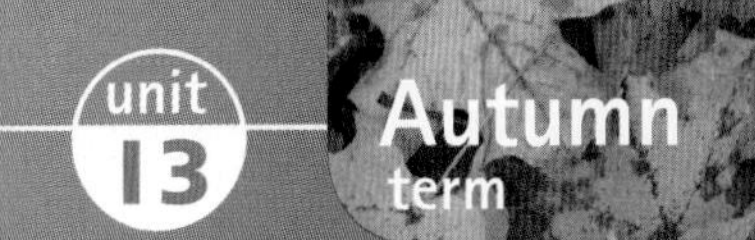

Who stole the painting?

A **A famous painting has been stolen from Lord Avalot's castle in Edinburgh. Use the clues to catch the thief.**

Clues

1. A footprint was found outside the window from a size 7 shoe.
2. The thief needed to reach 220 cm to lift down the painting.
3. 36 visitors saw the thief run off with the painting.
4. From the footprint, police estimate that the thief weighed at least 70 kg.
5. At 2:15 p.m. on the day in question Annie was in Gullin, Hitch was in Kyle, Bif was in Trent, Joey was in Burnham and Bosh was in Mull.

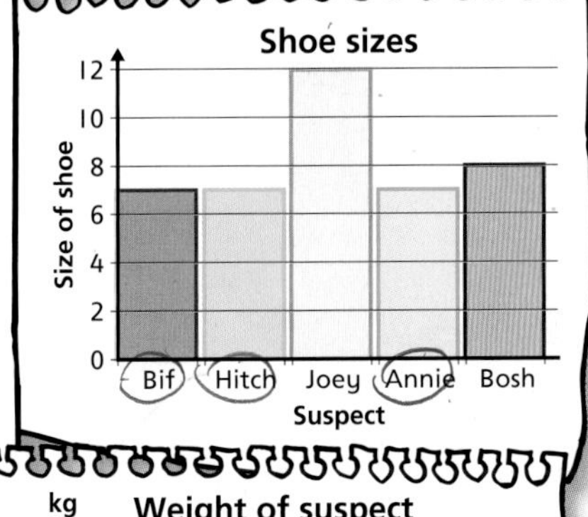

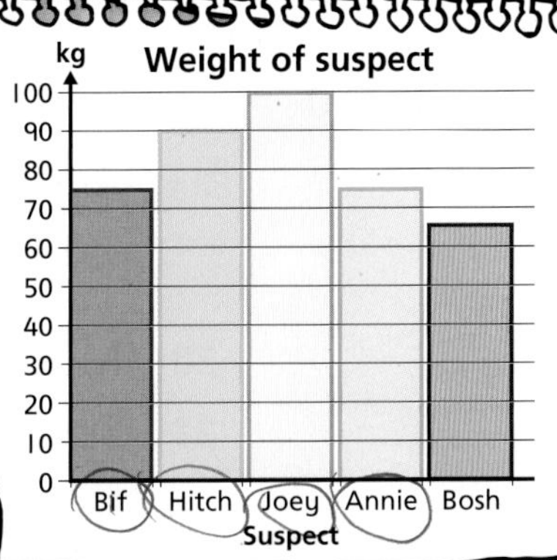

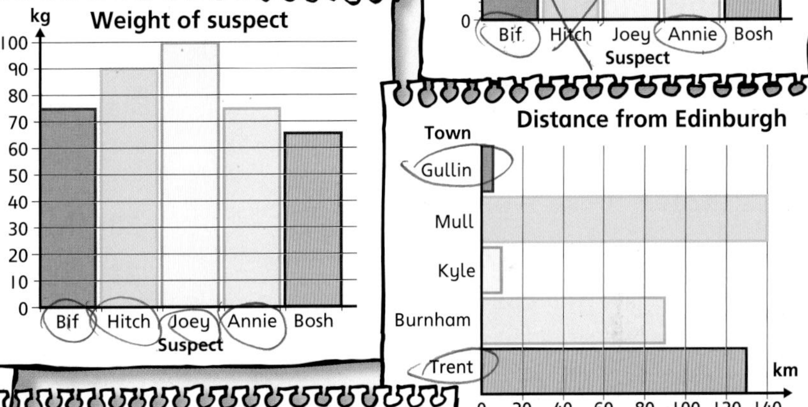

Time	Number of visitors
1:00 p.m. – 1:30 p.m.	25
1:30 p.m. – 2:00 p.m.	26
2:00 p.m. – 2:30 p.m.	43
2:30 p.m. – 3:00 p.m.	17

Challenge

Using the information in the bar charts above, make up your own clues to help identify a different thief.
Try drawing your own bar charts and make up clues for a different crime.

Review 2

A Copy and complete these sequences.

1 ■, 27, 21, ■, 9, 3
2 ■, 12, 20, ■, 36, ■
3 12, 4, ■, ⁻12, ■, ⁻28
4 16, ■, 4, ■, ⁻8, ⁻14

B Answer these.

5 36 × 4
6 67 × 3
7 $\frac{1}{2}$ of £360
8 $\frac{1}{4}$ of 72p
9 How many each and how many left over when 46 flies are shared equally among 3 spiders?

C Solve these word problems.

10 Jars of jam are packed in boxes of 8. If 1 jar costs £1·25, what is the cost of 4 boxes?
11 Amy thinks of a number and subtracts 32. Her answer is 36. What was her number?
12 Ivor Bun buys 4 cakes costing £2·63 each. How much change will he get from £20?

D Copy and complete.

13 $\frac{3}{5} = \frac{■}{10}$
14 $\frac{6}{8} = \frac{3}{■}$
15 $\frac{2}{2} = \frac{3}{■}$
16 $\frac{■}{3} = \frac{2}{6}$
17 Which is the greater amount: $\frac{7}{10}$ of £120, or $\frac{4}{5}$ of £100?

E Write how long these evening journeys take.

18

Start

Finish

19

Start

Finish

F Find the distance from Alton to each place.

20 Moulton
21 Soltown
22 Bolton
23 Colton

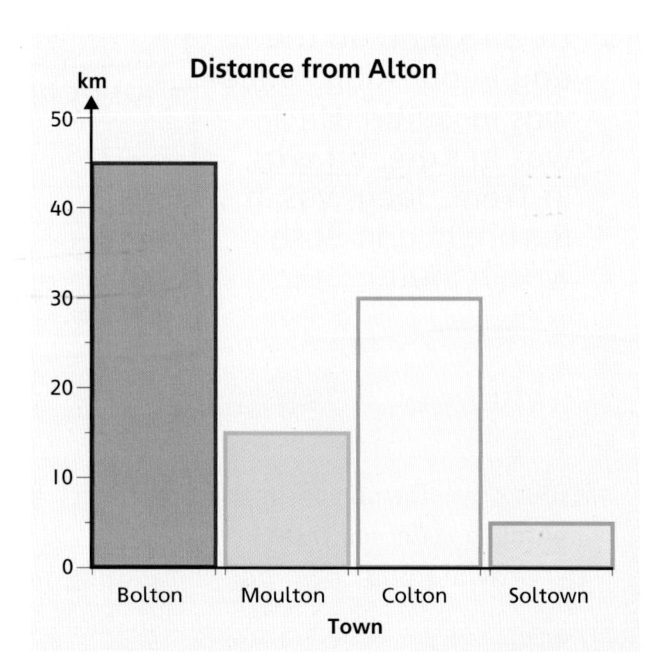

A Use each set of three numbers to make the largest and smallest number possible.

1 6 4 7
2 5 1 9
3 6 3 8
4 9 2 4
5 3 2 5
6 8 9 8
7 1 3 5

B Round these to the nearest 10.

8 40

8 36
9 27
10 41
11 86
12 25

C Round these to the nearest 100.

13 353
14 137
15 888
16 462
17 550

D Multiply each number by 10 and 100.

18 $63 \times 10 = 630$
$63 \times 100 = 6300$

18 63
19 47
20 15
21 68
22 29
23 93
24 33
25 86
26 3
27 74
28 57

E Multiply each number by 10.

29 326
30 85
31 164
32 295
33 902
34 610
35 40

F Divide each number by 10.

36 490
37 370
38 8260
39 2940
40 1950
41 7620
42 3000

G Multiply each number by 100.

43 27
44 5
45 56
46 32
47 89
48 164
49 439

Challenge

Place eight red and eight blue counters in a bag. Pick a ball below, close your eyes and take a counter from the bag.

5500	140	870	6200
490	620	6200	140
6120	61200	550	620
1400	8700	6120	4900

If the counter is blue, multiply your ball number by 10. If the counter is red, multiply your ball number by 100.

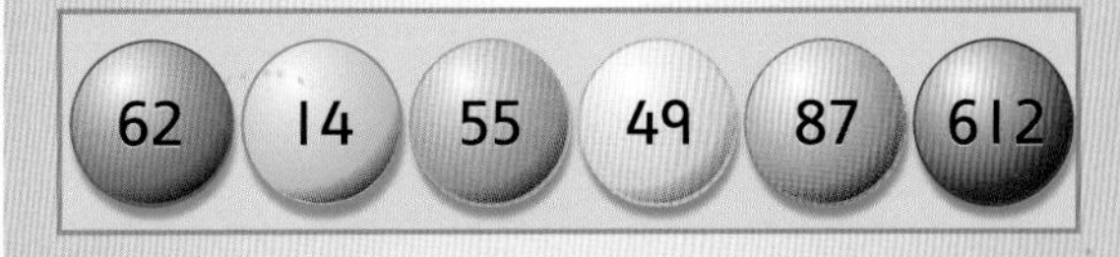

If your answer is on the grid, cover it with a red counter from the bag.
Your partner takes a turn using blue counters.
Continue until one player makes a line of three counters. That player wins.

Can you multiply a 3-digit number by 10 and divide a 4-digit number by 10?

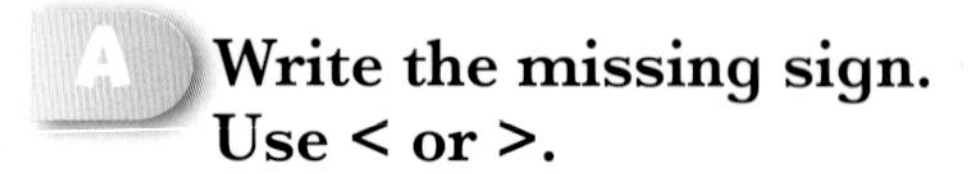

A Write the missing sign.
Use < or >.

1 4163 ★ 3614
2 2496 ■ 2649
3 8103 ● 8013
4 6904 ● 6409
5 3060 ■ 3006
6 7491 ◖ 7194
7 8866 ▲ 8686

1 4163 > 3614

B Use each set of four digits to make eight different numbers. Write the eight numbers in order, smallest first.

8 6 9 8 1
9 5 3 6 1
10 9 1 4 3
11 8 1 0 5
12 6 6 7 8

8 1689, 1968, 6819, 6981, 8961, 9168, 9186, 9816

C Use number cards 1 to 9. Shuffle the cards. Turn over four cards and place them to make each diagram correct. Record the result. Repeat four times for each diagram.

13 ▭ ▭ > ▭ ▭

14 ▭ ▭ < ▭ ▭

Use the top six cards for these diagrams.

15 ▭ ▭ ▭ < ▭ ▭ ▭

16 ▭ ▭ ▭ > ▭ ▭ ▭

Challenge

Use number cards 0 to 9. Shuffle the cards and turn over the top four. Use the cards to make a number that fits in the blue section of the number line. If this is not possible, write 'Blue impossible'. Repeat for the yellow and red sections of the number line. Record your results. Take five more turns.

9 1 8 7
Blue 1897
Yellow impossible
Red 8971

1000 — 3000 — 6000 — 10 000

A Round these heights to the nearest 10 cm.

1 168 cm
2 143 cm
3 186 cm
4 158 cm
5 191 cm
6 175 cm

1 170 cm

B Round the distance from Manchester of each town to the nearest 10 km.

7 London 322 km
8 Plymouth 449 km
9 Norwich 283 km
10 Aberdeen 514 km
11 Holyhead 192 km
12 Exeter 385 km

C Round these weights to the nearest 100 g.

13 2920 g
14 2581 g
15 6295 g
16 1850 g
17 4003 g

D Write the missing numbers on the number lines.

18 1694, 1697, 1699, 1700, 1701

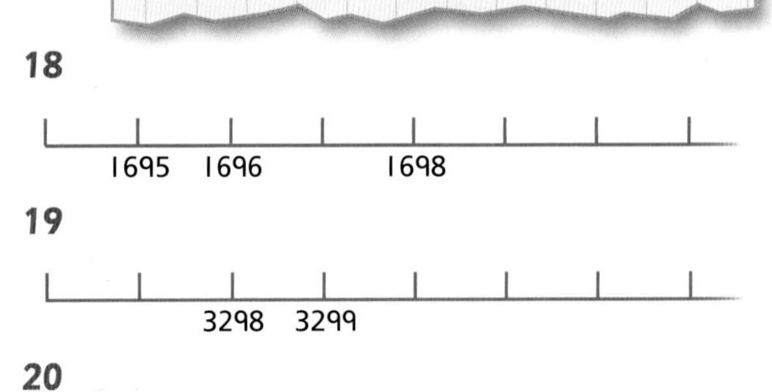

E Round each number to the nearest 10, 100 and 1000.

21 7658
22 9514
23 4659
24 8111
25 6235
26 7505
27 6350
28 5500

21 7660, 7700, 8000

Challenge

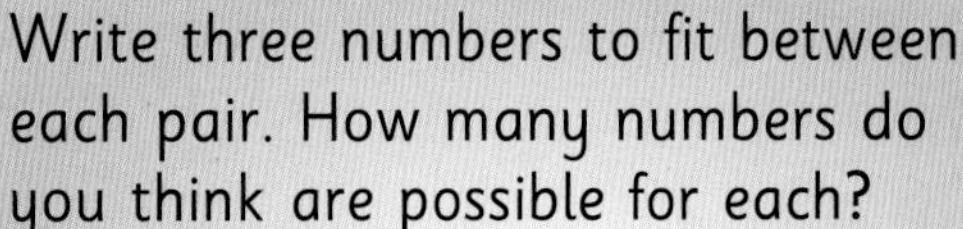

Write three numbers to fit between each pair. How many numbers do you think are possible for each?

a 623 > ★ > 495
b 379 > ■ > 299
c 897 < ● < 901
d 1694 < ● < 1700
e 8326 > ■ > 7900
f 2997 < ● < 3001

A Copy and complete these number squares.

1

+	6	4	9	7
3				
8				
9				
10				

2

+				
6	15			
8			13	
5		11		
9				16

B Write 'true' or 'false' for each.

3 $6 + 9 = 9 + 6$
4 $6 + 5 + 4 = 4 + 6 + 5$
5 $26 + 25 + 14 = 40 + 25$
6 $26 - 9 > 26$
7 $1001 - 11 < 1001$
8 $28 + 33 = 50 + 8 + 3$
9 $19 + 11 + 17 = 30 + 19$
10 $36 - 18 = 18 - 36$
11 $8 + 3 + 2 + 7 = 10 + 10$
12 $30 + 50 + 70 = 100 + 50$
13 $40 + 70 + 60 = 170$

LIE DETECTOR

C Answer these.

14 $58 + 19$
15 $26 + 11$
16 $47 + 9$
17 $75 + 19$
18 $136 + 31$
19 $226 + 29$
20 $47 + 49$
21 $124 - 29$
22 $861 - 59$
23 $87 + 71$

Challenge

Copy each grid and shade a route to show how to move from the first number to the second number.

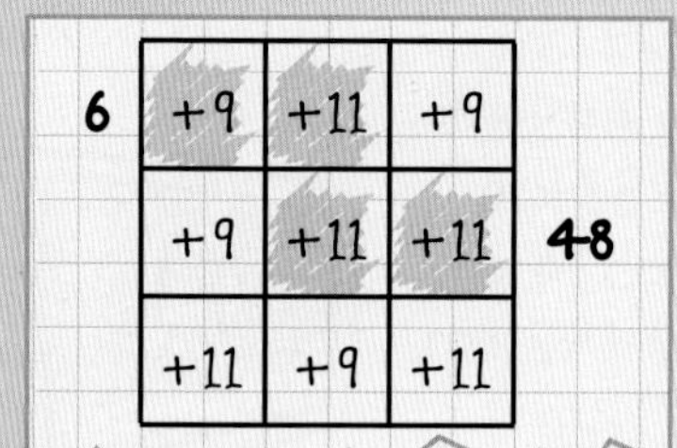

6 + 9 + 11 + 11 + 11 = 48

a

	+11	+9	+11	58
	+11	+9	+9	
9	+9	+9	+11	

b

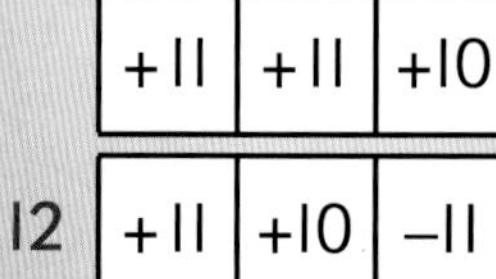

	+9	+10	+9	69
8	+11	+9	+9	
	+11	+11	+10	

c

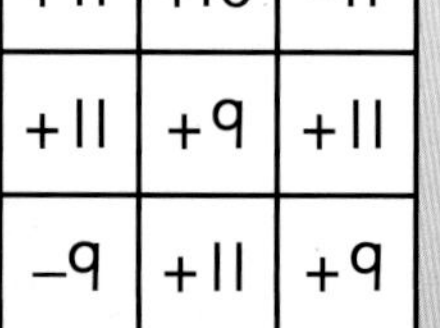

12	+11	+10	−11	
	+11	+9	+11	
	−9	+11	+9	45

Can you add or subtract numbers by rounding to the nearest multiple of 10 and adjusting your answer?

A Solve these word problems.

1 A fish weighs 1226 g. A kitten weighs 29 g more than the fish. How heavy is the kitten?

2 A golden eagle is 59 cm longer than a starling. How long is the starling if the eagle measures 82 cm?

3 Lincoln is 19 km further from Newcastle than Derby. If Lincoln is 245 km from Newcastle, how far is Derby from Newcastle?

4 Edmund climbs 1326 m. Hilary climbs 191 m more. How high does Hilary climb?

5 In the football league, Rovers have scored 39 less goals than Rangers. Rangers have scored 72 goals. How many goals have Rovers scored?

6 In four throws at darts, Bill scores 26, 29, 41 and 54. What is his total score?

7 Ivor Lot has £429 more than Nealy Rich. If Ivor has £864, how much has Nealy?

8 Jim eats 420 cm of spaghetti. Jerry eats 510 cm. Lisa eats 370 cm. What length of spaghetti do they eat altogether?

9 Mrs Baxter wants to buy this car. She has saved £6200. How much more does she need?

£9600

10 Hugo First and Hugo Second both give some money to charity. Hugo First gives £3800. The charity receives £5600 from them in total. How much does Hugo Second give?

Challenge

Find the missing numbers in these calculations. Use a calculator to add together all the missing numbers. Which colour balloon shows your result? If none do, try again!

a 52 + 40 = ★
b 320 + 36 = ■
c 640 + ● = 682
d 69 + ● = 100
e 789 + ■ = 800
f 6200 + ● = 7000
g 800 + 500 = ▲
h 846 – 29 = ●
i 790 + 11 = ●

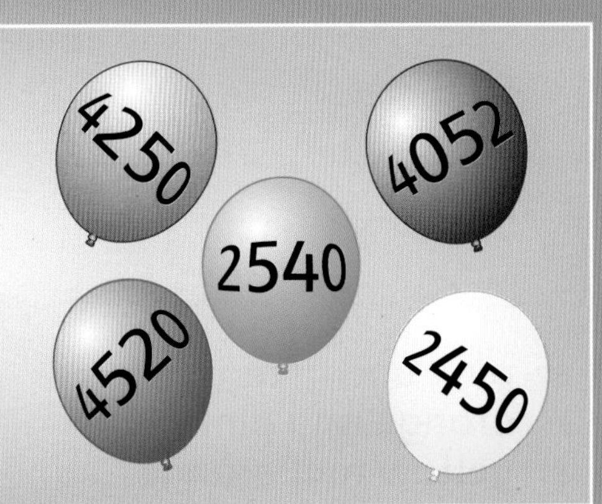

Can you solve problems by rounding to the nearest multiple of 10 and adjusting your answer?

A Use partitioning to find the total of each pair of cheques.

1 £45 + £37
  40 + 5
+ 30 + 7
  70 + 12 = 82
Total = £82

1 blue and green
2 green and orange
3 pink and orange
4 pink and green
5 yellow and blue

B Use partitioning to work out the difference in value between each pair of cheques.

6 £56 − £37

  50 + 6
− 30 + 7

  40 + 16
− 30 + 7
  10 + 9 = 19
Difference = £19

6 pink and green
7 orange and pink
8 yellow and orange
9 blue and yellow

C Use partitioning to complete these.

10 362 + 149
11 427 + 169
12 329 + 181
13 621 + 193
14 338 + 187
15 742 + 179
16 246 + 642
17 486 + 275

10 362 + 149
  300 + 60 + 2
+ 100 + 40 + 9
  400 + 100 + 11 = 511

Challenge

Use a 100 square. Shade squares to form two of your initials. Your first initial must include the number 63. Your second must include the number 99.

1	2	3	4	5	6	7	8	9	10
11	12	13	14	15	16	17	18	19	20
21	22	23	24	25	26	27	28	29	30
31	32	33	34	35	36	37	38	39	40
41	42	43	44	45	46	47	48	49	50
51	52	53	54	55	56	57	58	59	60
61	62	63	64	65	66	67	68	69	70
71	72	73	74	75	76	77	78	79	80
81	82	83	84	85	86	87	88	89	90
91	92	93	94	95	96	97	98	99	100

Choose one number from each initial and add them together. Repeat to find as many answers as you can.

Can you add or subtract mentally any pair of 2-digit numbers?

A Answer these. If you use a written method, show your working.

1 62 + 39
2 37 + 22
3 45 + 37
4 19 + 48
5 28 + 86
6 42 + 79
7 323 + 116
8 425 + 47
9 163 + 498

B Answer these. If you use a written method, show your working.

10 83 − 22
11 59 − 46
12 59 − 6
13 76 − 22
14 41 − 23
15 82 − 37
16 64 − 19
17 93 − 38

C Use a written method you know to find the total of these scores.

18 2 red rings
19 2 blue rings
20 2 yellow rings
21 2 green rings
22 black ring and pink ring

D Use a written method you know to find the difference between these scores.

23 black ring and orange ring
24 brown ring and black ring
25 orange ring and pink ring
26 brown ring and orange ring

Challenge

Look at the hoop-la game.
a Find three rings that score a total of 630. Write their colours.
b Find two rings that give a difference of 278. Write their colours.

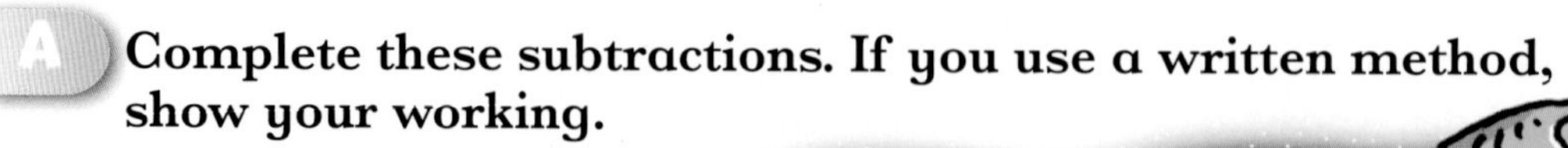

A Complete these subtractions. If you use a written method, show your working.

1 149

1 235 − 86
2 347 − 139
3 270 − 96
4 513 − 184
5 205 − 79
6 817 − 119
7 742 − 187
8 952 − 386
9 562 − 278
10 602 − 186
11 419 − 189
12 702 − 345
13 622 − 177
14 526 − 278
15 311 − 94
16 506 − 348

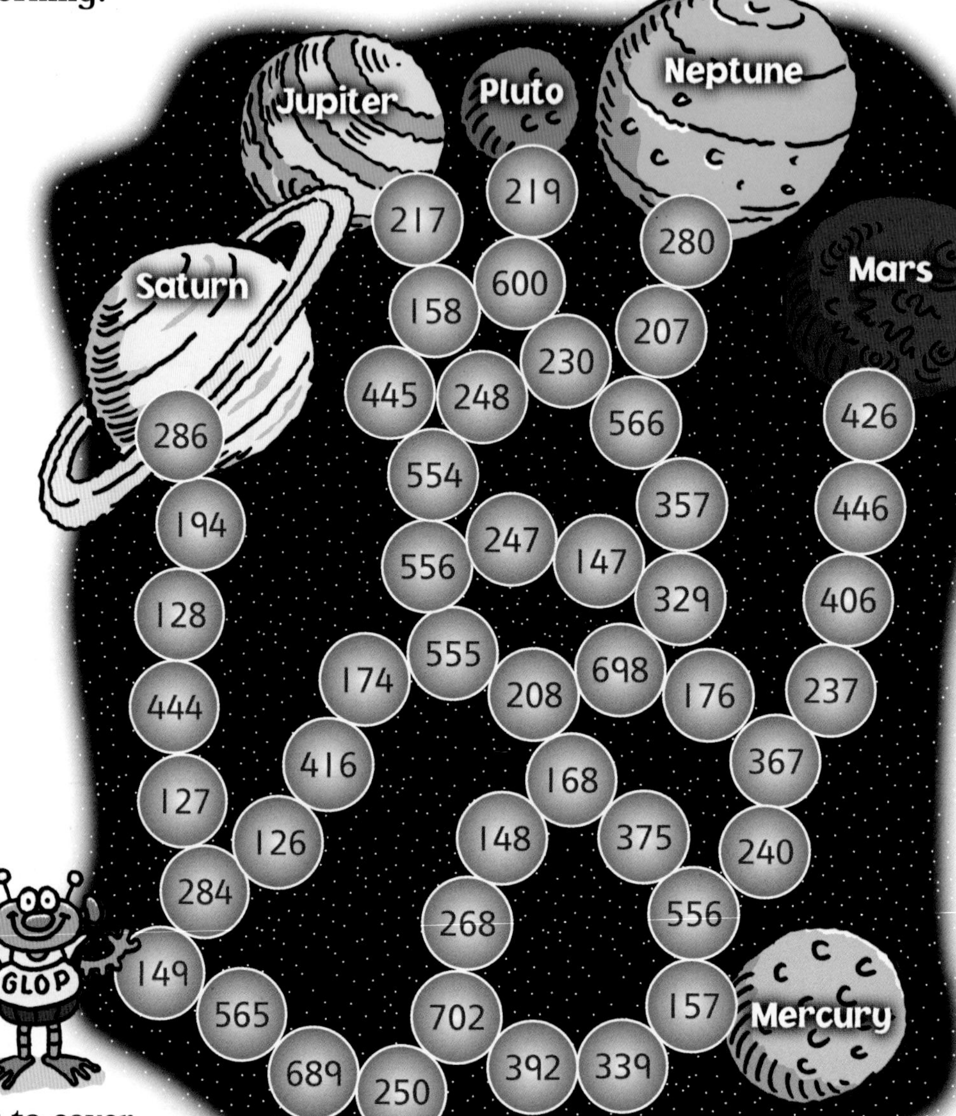

B Use counters to cover numbers on Glop's space route to match the answer to each subtraction above.

17 Write the numbers of Glop's route through space in the order he passes them.
18 Write the name of the planet Glop visits.

Challenge

Work out a set of subtraction clues to take Glop to a different planet.

Can you use a written method to subtract 2- or 3-digit numbers?

A Write how much each person pays for their meal.

EASTERN SPICE *Indian Resturant*	
Papadom	£0·48
Samosa	£2·50
Onion Bhaji	£2·25
Prawn Puree	£3·30
Tandoori Chicken	£5·25
Chicken Tikka Masala	£6·85
Vegetable Balti	£5·75
Lamb Biriani	£7·45
Pilau Rice	£1·65
Nan Bread	£1·95
Ice-cream	£1·72

1 Eta Lot

Papadom
Tandoori Chicken
Ice-cream

1 £0·48
£5·25
\+ £1·72
£7·45

2 Ivor Plateful

Samosa
Chicken Tikka Masala
Nan Bread

3 Tom Ahtoe

Papadom
Onion Bhaji
Vegetable Balti

4 Betty Likesit

Prawn Puree
Lamb Biriani
Ice-cream

5 Miss Derchips

2 Papadoms
Tandoori Chicken
Pilau Rice

B Find the difference in the amounts spent by these people.

6 Ivor and Miss Derchips

7 Betty and Tom

8 Ivor and Tom

9 Miss Derchips and Betty

6 £3·44

C Find how much money was spent altogether at the Eastern Spice Restaurant in these months.

10 January and February

11 February and March

12 April and May

13 January, February and March

14 March, April and May

Money spent at Eastern Spice Restaurant		
January	£	8265
February	£	7827
March	£	9694
April	£	4688
May	£	7557

Challenge

Plan four different meals from the Eastern Spice menu. Work out the cost of each meal.

You need:

- two sets of coloured counters
- a dice
- a partner

Each place one counter on START. Take turns to throw the dice and move your counter round the game board.

Answer the question you land on and find the answer on a hexagon in the centre. Cover it with a counter. If your answer is not showing, miss a turn.

The first player to link their zones with a pathway of counters is the winner.

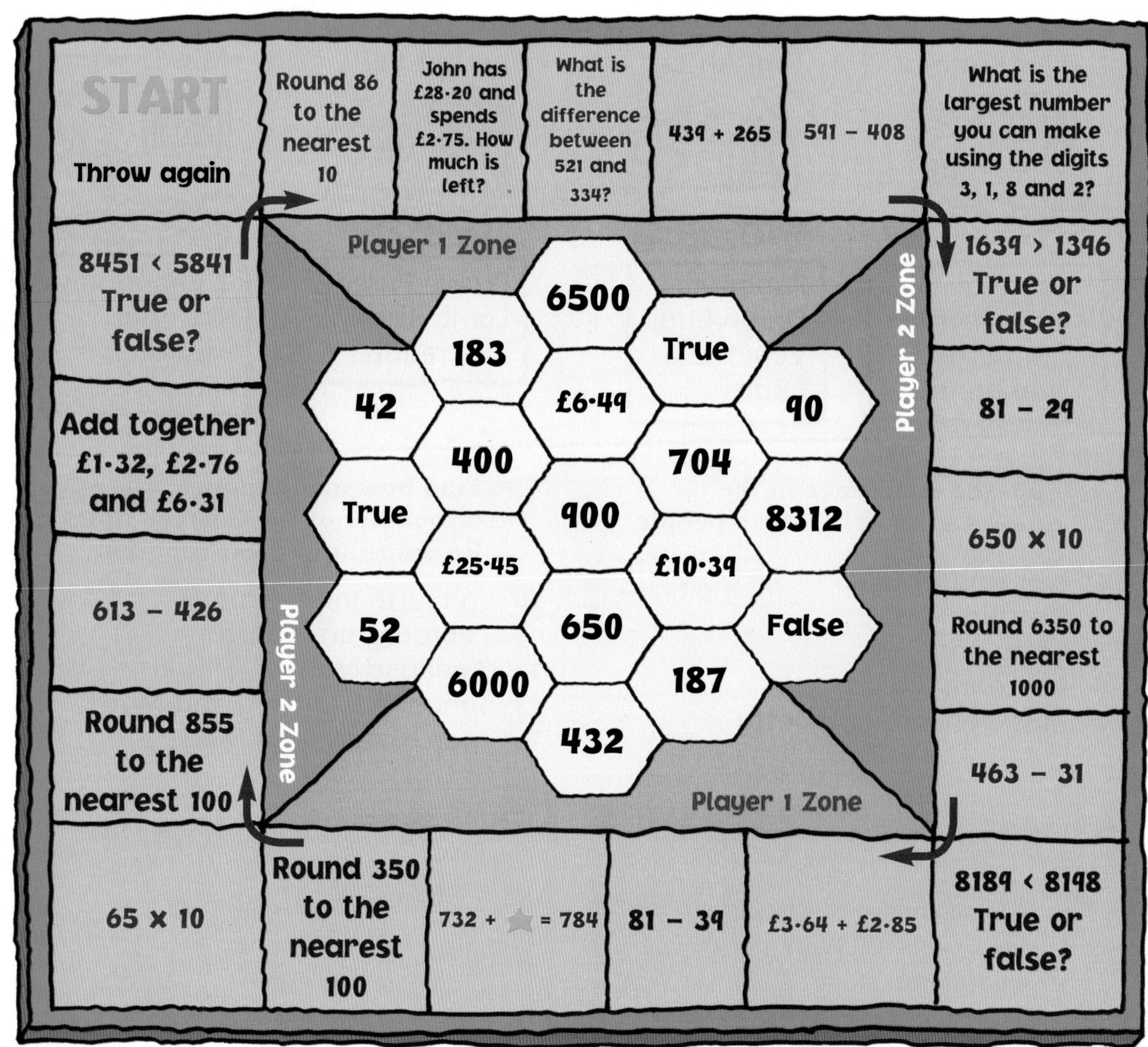

A Answer these time questions.

1 How many days in a week?
2 How many months in a year?
3 How many hours in a day?
4 How many seconds in a minute?
5 How many days in a year?
6 How many full weeks in a year?
7 What is today's date?
8 What is your date of birth? Give the day, month and year.

B Write each time.

9 9:40 or
twenty to ten

9

11

10

12

C Write in hours and minutes.

13 1 hour
25 minutes

13 85 minutes
14 79 minutes
15 100 minutes
16 124 minutes
17 99 minutes
18 136 minutes
19 180 minutes
20 162 minutes

D Write in minutes and seconds.

21 98 seconds
22 65 seconds
23 87 seconds
24 101 seconds
25 122 seconds
26 154 seconds
27 200 seconds
28 111 seconds
29 211 seconds

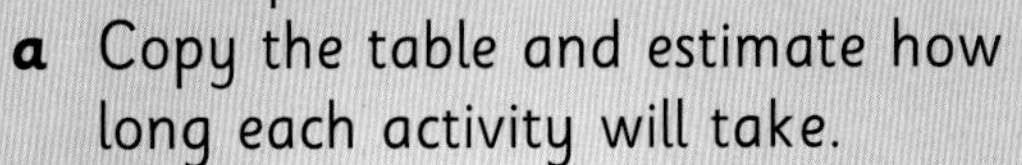

Challenge

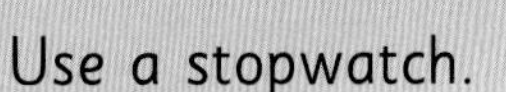

Use a stopwatch.
a Copy the table and estimate how long each activity will take.
b Time how long it takes to do each activity.
c Work out the differences between your estimates and measures to complete the table.

Activity	Times (to nearest second)		
	Estimate	Measure	Difference
Count to 100			
Blink 30 times			
Write your name and address			
Tap your finger 50 times			
Write the alphabet			

A Choose the best estimate for the mass of each object.

1

40 g 400 g 40 kg

2

4 g 40 g 400 g

3

3 g 30 g 300 g

4

50 g $\frac{1}{2}$ kg 5 kg

5

50 g $\frac{1}{2}$ kg 50 kg

6

50 g 500 g 5 g

B Write in grams.

7 $\frac{1}{2}$ kg
8 $\frac{1}{4}$ kg
9 $\frac{3}{4}$ kg
10 $\frac{1}{10}$ kg
11 $\frac{7}{10}$ kg
12 2 kg

C Write which mass is heavier.

13 2 kg or 1975 g
14 $\frac{1}{2}$ kg or 475 g
15 936 g or $\frac{9}{10}$ kg
16 $\frac{3}{4}$ kg or 700 g

D Write the approximate measurement shown on each set of scales.

17
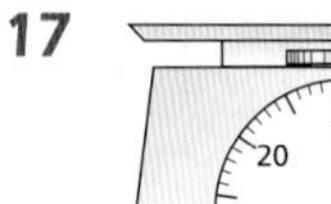
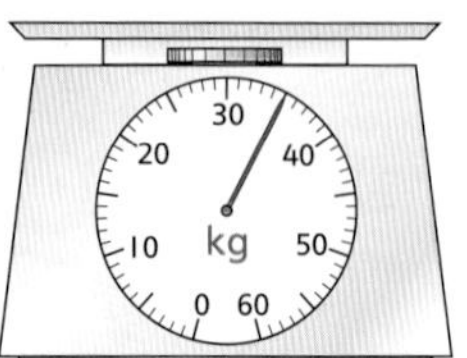

18
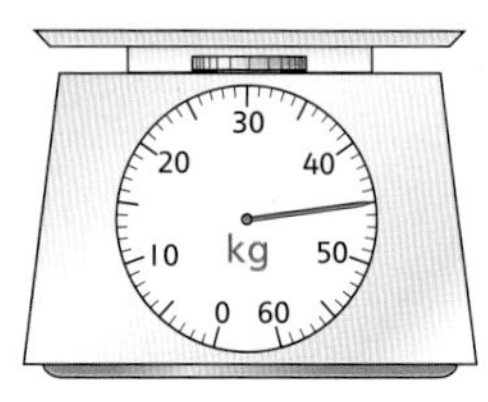

20
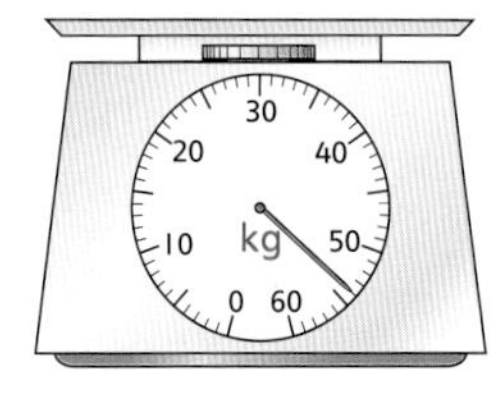

19
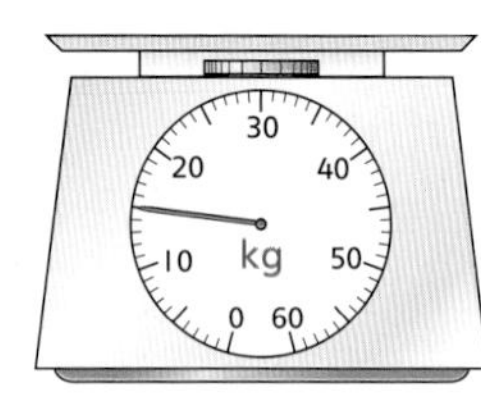

21
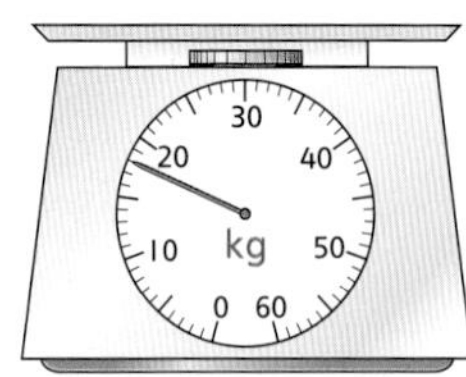

Challenge

Work with a partner. Use a set of scales.

a Each collect five objects that you think have a mass of less than 1 kg.
b Estimate the mass of each object to the nearest 10 g.
c Weigh each object to the nearest 10 g.
d Work out the difference between each estimate and mass.
e Add together all five differences.

The person with the lower total wins.

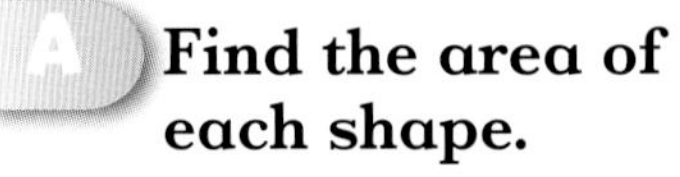

A Find the area of each shape.

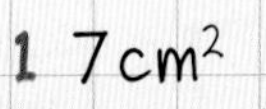

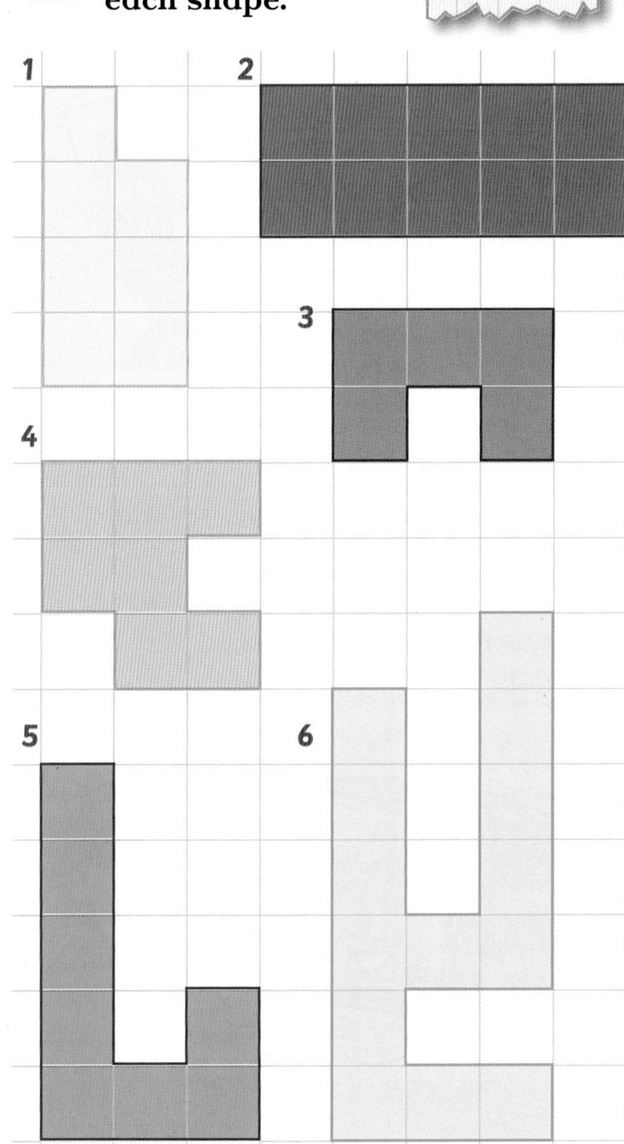

B Work out the perimeter of these shapes.

7 yellow shape
8 red shape
9 purple shape
10 orange shape
11 pink shape

C Write the area of each shape.

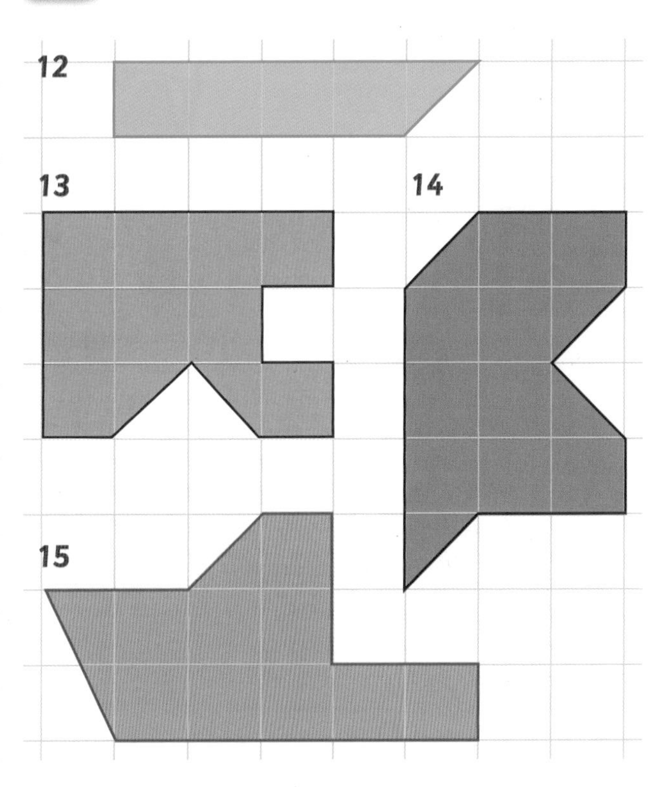

D Use cm-squared paper. Draw three different shapes for each of these. Write the area and perimeter for each shape.

16 perimeter 22 cm
17 area 16 cm^2

Challenge

a What is the largest area you can cover using a rectangle with a perimeter of 18 cm?
b Use cm-squared paper. Draw a castle to cover an area of 100 cm^2.

Can you draw different rectangles with the same perimeter and find their area?

A Write the missing sign in each calculation.

1 62 ★ 24 = 86
2 71 ■ 46 = 25
3 51 ✿ 42 = 93
4 120 ● 10 = 12
5 9 ■ 5 = 45
6 8 ◖ 6 = 48
7 32 ▲ 8 = 4
8 49 ● 7 = 7
9 129 ● 64 = 65

B Ivor Problem has emailed the online Homework Helpdesk for help with some maths questions. Reply to Ivor and help him to understand each question.

10 Ivor, you can do this in your head. Take away 20 from 46. This leaves 26, but you have taken away 1 too many, so add it back on. That gives you 27.

10 46 − 19
11 35 + 16
12 57 − 19
13 Tickets for the cinema cost £4·60. What is the cost of 4 tickets?
14 A mobile phone costs £56·50. A top-up card is £15·00. What is the total?
15 457 + 265

Challenge

Ivor has finished his homework, but he is worried about his answers to questions 13, 14 and 15. Write an email to tell Ivor a good way to check each answer.

Can you explain and record how you carried out a calculation?

A Make up a problem to match each calculation.

1. $124 - 29 = 95$
2. $210 - 63 = 147$
3. $47 + 68 = 115$
4. $6 \times 9 = 54$
5. $72 \div 4 = 18$
6. $268 + 732 = 1000$
7. $60 \times 5 = 300$

> 1 During the football season United scored 124 goals. If Brian Griggs scored 29 goals, how many goals were scored by the rest of the team?

B Solve each word problem. Explain briefly what you do.

> 8 1 bun costs half of 64p = 32p
> 5 buns cost 5 lots of 32p
> 5 lots of 30 p = 150p = £1·50
> 5 lots of 2p = 10p
> 5 buns cost £1·50 + 10p = £1·60

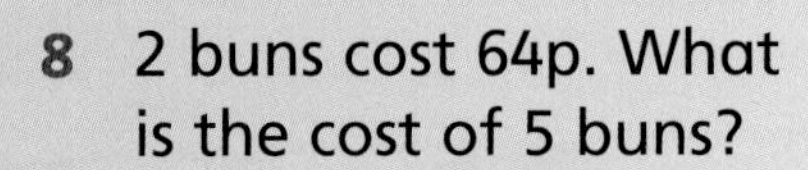

8. 2 buns cost 64p. What is the cost of 5 buns?
9. How many children can each have 4 sweets from a packet of 48 sweets?
10. For a concert, chairs are set out with 18 in each row. How many chairs in 6 rows?

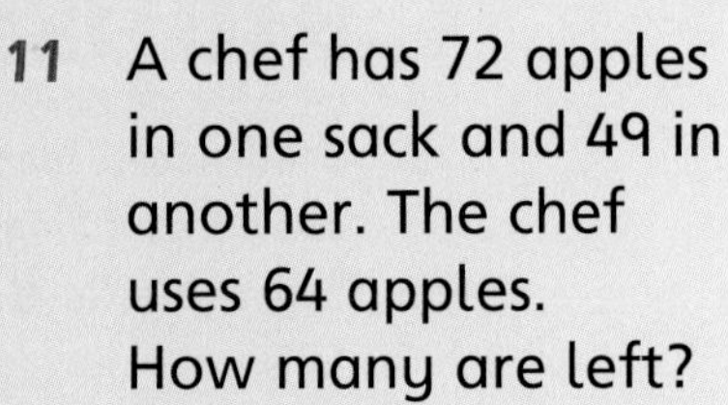

11. A chef has 72 apples in one sack and 49 in another. The chef uses 64 apples. How many are left?
12. In a sponsored walk 4 children walk 15 km and 1 child walks 19 km. How many kilometres do they walk altogether?
13. There are 11 cherries on a pie. How many pies can be made with 200 cherries? How many cherries are left over?

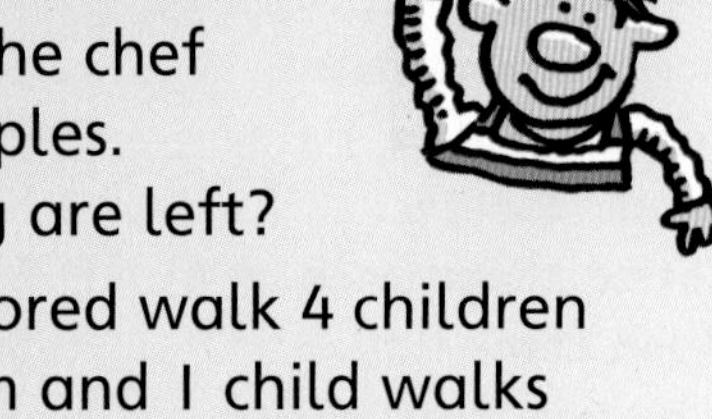

Challenge

a 28 → 126 Enter 28 in your calculator. Using only the keys 2, 3, +, × and =, can you reach the number 126?

b 4 → 199 Can you go from 4 to 199 on your calculator using only the keys 5, 2, +, × and =? You must only press 20 keys or less.

Can you choose and use appropriate operations to solve word problems?

A **You are going camping with 9 friends for 2 days. You have £100 to spend on food.**

1. Plan your menu for all 10 people from the list. Make sure you take enough for everyone!
2. Work out the total cost of your food.
3. Write how much change you have left from £100.

Campsite shop

Item	Price	Item	Price	Item	Price
Corn Flakes	£1·78	Sandwiches (pack of 2, choice of fillings)	£1·69	Milk (1 litre)	58p
Honey Crunch Flakes	£1·82	Sausage rolls (for 8)	£2·00	Sugar (1 kg)	92p
Eggs (box of 6)	£0·45	Cola	£0·88	Pizza slice	£0·22
Bacon (8 slices)	£1·40	Lemonade	£0·72	Chicken nuggets (bag of 25)	£1·56
Cheese slices (pack of 8)	£1·59	Orange squash	£1·45	Baked beans	£0·46
Bread (large loaf)	£0·80	Packet of biscuits	£0·45	Butter (500 g)	£0·93
Tomatoes (pack of 20)	£1·46	Cake	£0·40	Samosa	£0·34
Jam (large jar)	£1·40	Choc ices (for 10)	£4·80	Fish fingers (box of 16)	£2·08
Marmalade (large jar)	£1·05	Rolls (for 4)	88p	Burgers (for 4)	£3·20
Tomato sauce (large bottle)	£0·75	Trifle (each)	35p	Sausages (for 8)	£1·00
Crisps (packet of 4 bags)	£1·20			Vegetable burgers (for 4)	£3·00

B **These are the other costs for the camp. Answer these questions.**

Transport £65 First aid £8·48 Camp site fees £25 Equipment £28·78

4. What is the total cost of the camp for all 10 children?
5. How much will each person need to pay?

Challenge

Use the digits 1, 2, 3 and 4 at least once and any of the operations +, –, × and ÷. Make a number sentence to give each answer.

a 30 **b** 22 **c** 63 **d** 96 **e** 250 ***f*** 114

Can you use appropriate strategies to solve open-ended problems?

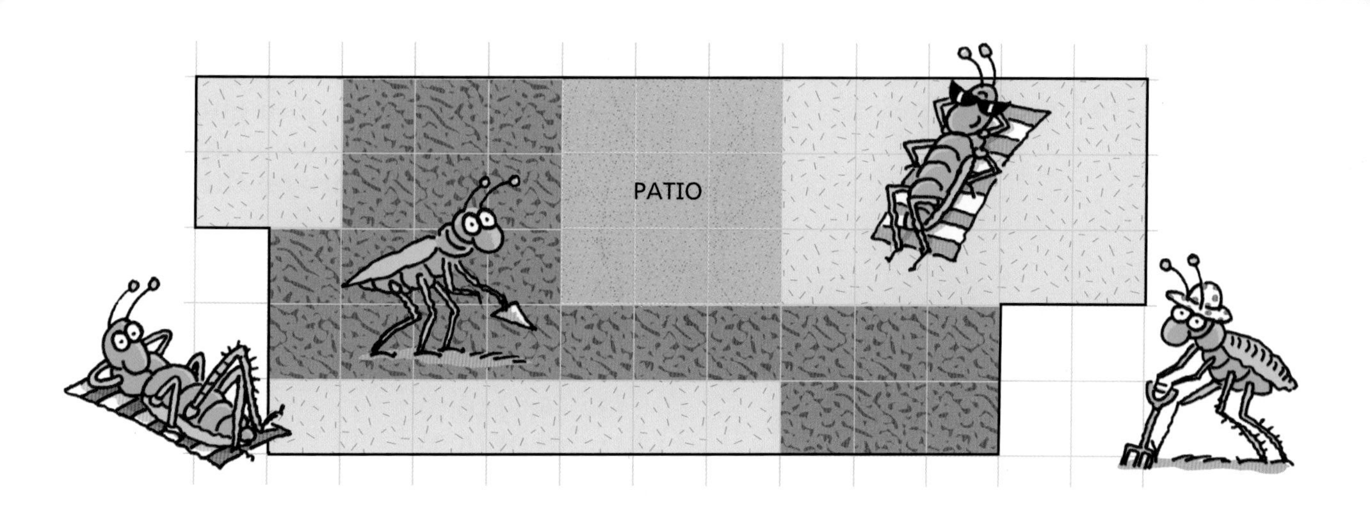

Look at the plan of the grasshoppers' garden. Each square is 1 cm by 1 cm. Answer these questions.

1 What is the total area of the grasshoppers' garden?
2 What is the area of the patio?
3 The grasshoppers enjoy hopping around the perimeter of their garden. How far is that?
4 The brown areas are for planting vegetables. The mass of seeds needed for each square centimetre is 15 g. What mass of seeds do the grasshoppers need altogether?
5 The mass of each patio slab is 135 g. What is the total mass of all the patio slabs?
6 The green areas are for grass. Grass seeds cost £0·13 per square centimetre. What is the cost of planting the grass?
7 How long was each grasshopper gardening?

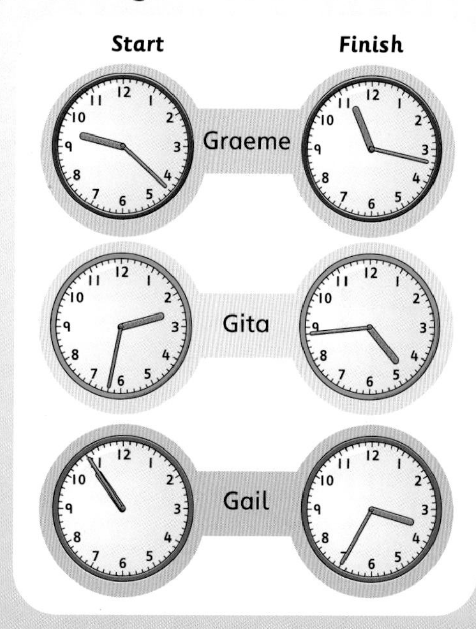

8 Every day it takes a grasshopper 10 minutes to clean 1 patio slab, 25 minutes to water 1 cm^2 of grass, and half an hour to dig 1 cm^2 of the vegetable area. How much time do the grasshoppers spend looking after the garden:

a in 1 day?
b in 1 week?
c in June?

Direction and angle

A Write the letter of the smallest angle for each shape.

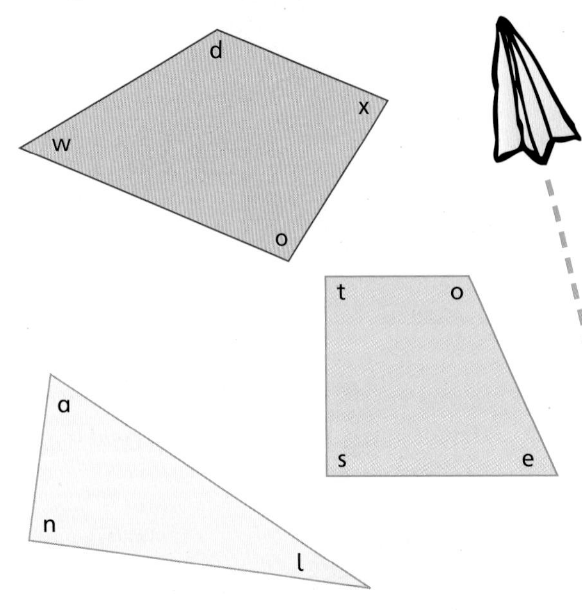

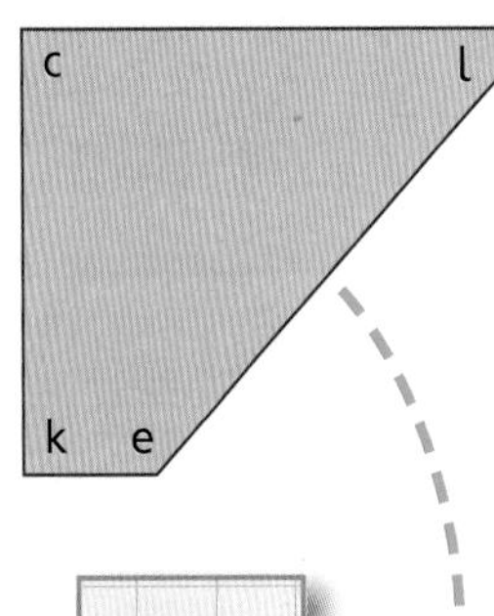

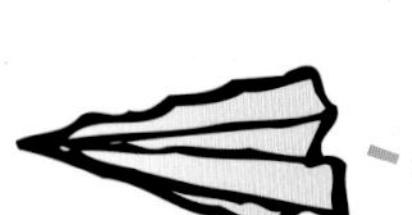

1 w

1 red shape
2 green shape
3 yellow shape
4 purple shape

B Write the letter of the largest angle for each shape.

5 red shape
6 green shape
7 yellow shape
8 purple shape

Use the letters from your answers to find a hidden message.

C On a co-ordinate grid draw a straight line to join each pair of points. Write the co-ordinates of two other points on the line.

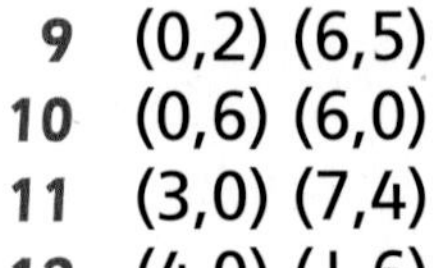

9 (0,2) (6,5)
10 (0,6) (6,0)
11 (3,0) (7,4)
12 (4,0) (1,6)

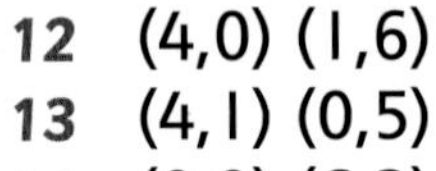

13 (4,1) (0,5)
14 (0,0) (6,3)

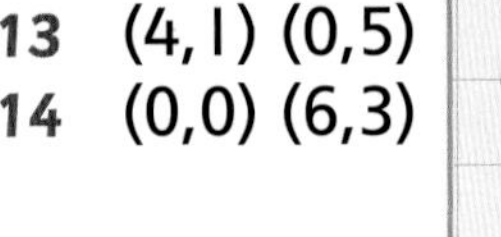

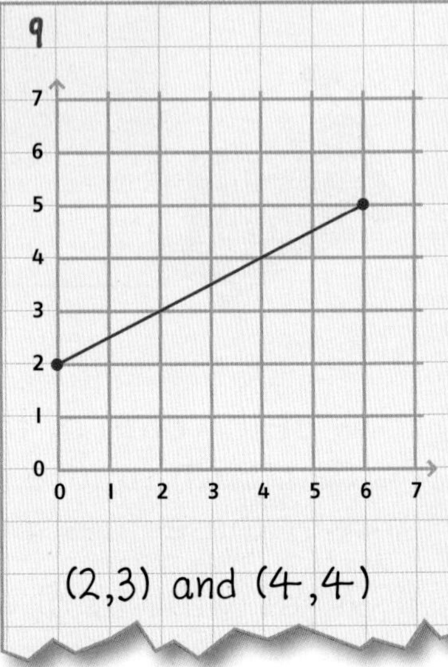

D Write the co-ordinates of two other points on the line when a horizontal line is drawn from each of these points.

15 (1,1)
16 (2,2)
17 (1,5)
18 (0,6)
19 (3,7)

15 (3,1) and (6,1)

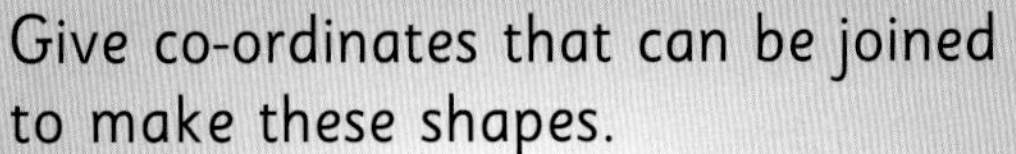

Challenge

Give co-ordinates that can be joined to make these shapes.

a rectangle
b square
c isosceles triangle
d right-angled triangle
e hexagon

Can you use co-ordinates to find the position of a point on a grid?

A Copy the diagram. Write the missing compass points.

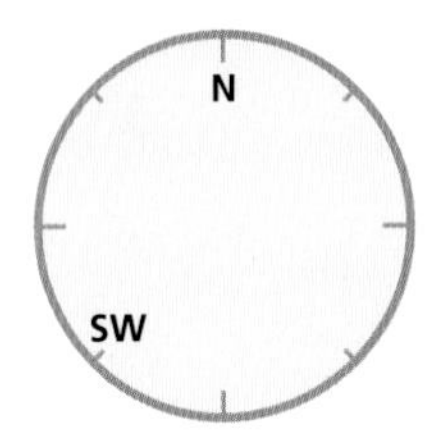

B Write the compass points in full.

1 NW
2 E
3 NE
4 SW
5 SE

1 north-west

C Find which direction the telescope will face after each turn.

	start facing	turn	new direction
6	E	$1\frac{1}{2}$ right angles anti-clockwise	NW
7	N	$\frac{1}{2}$ right angle clockwise	
8	S	1 right angle anti-clockwise	
9	SE	2 right angles clockwise	
10	NE	$1\frac{1}{2}$ right angles clockwise	
11	SW	$2\frac{1}{2}$ right angles anti-clockwise	

D Write which direction the ship should sail to reach each point.

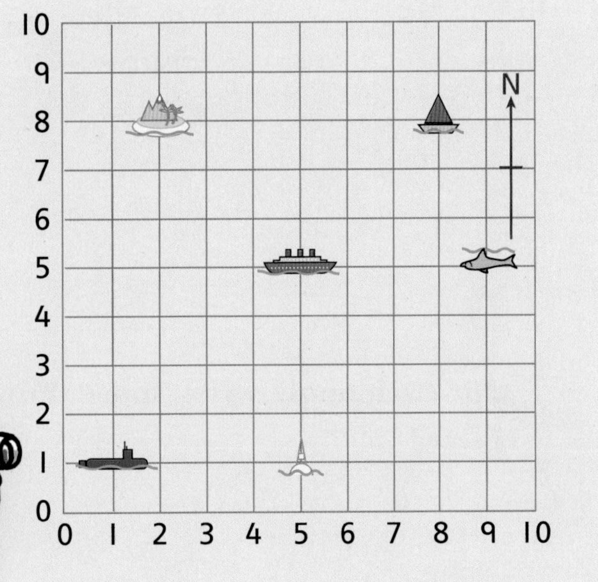

12 island
13 shark
14 buoy
15 point (10,0)
16 submarine
17 yacht

Challenge

Use directions to complete the instructions for drawing this spaceship.

Start with these instructions: 4 sides north, 1 diagonal north-east, 1 side west, ...

Now write instructions for your own drawing.

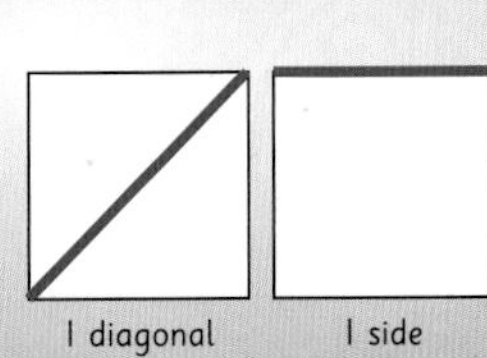

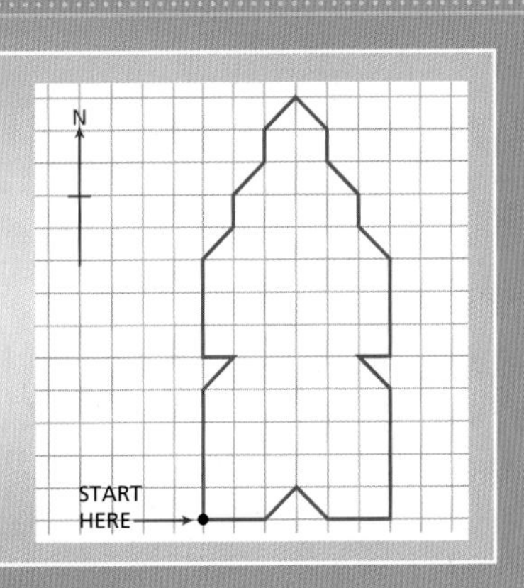

A For each angle write 'less than 90 degrees', 'greater than 90 degrees', or 'right angle'.

1 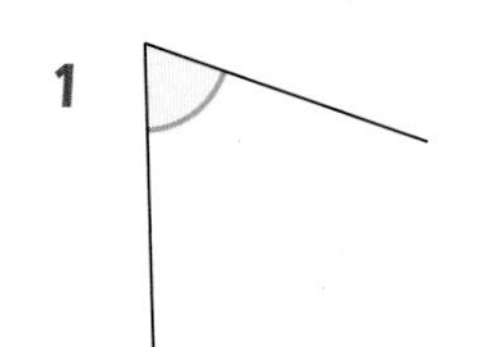

2 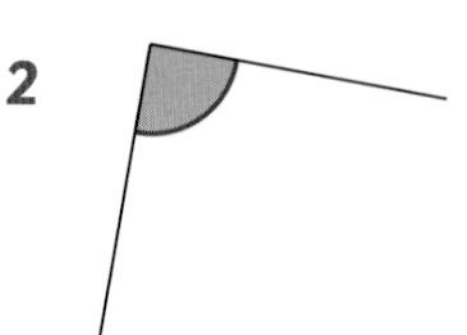

3 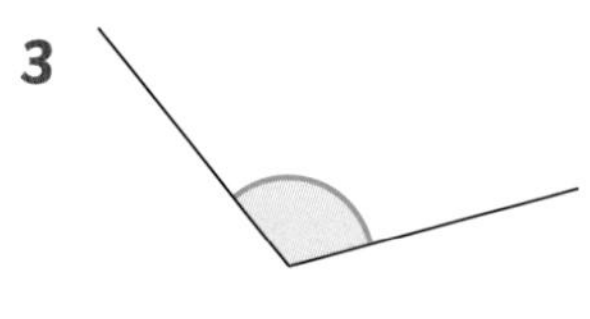

4 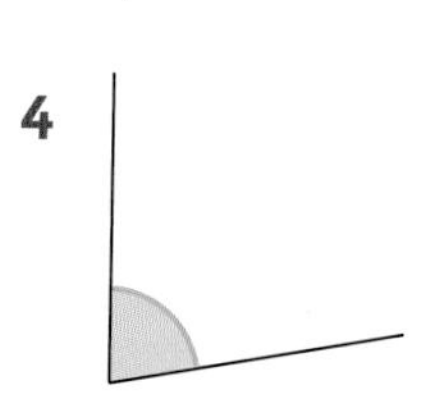

5 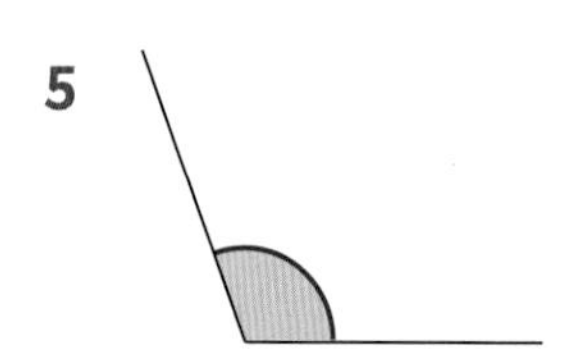

6 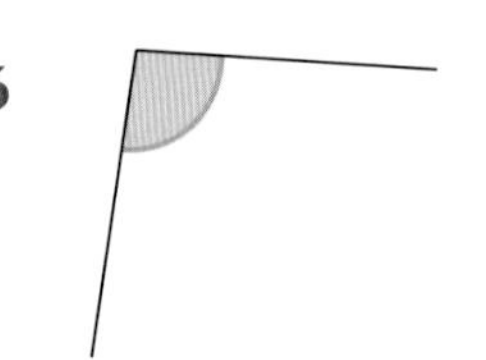

B For each angle write 'more than 45 degrees' or 'less than 45 degrees'.

7 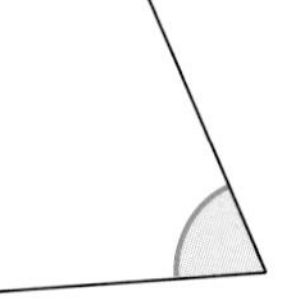

8 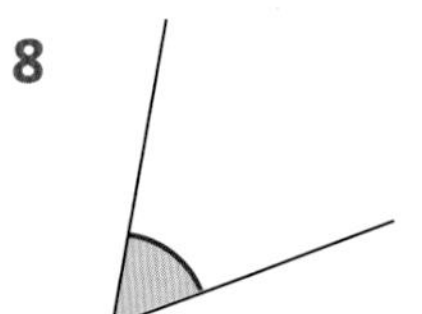

9 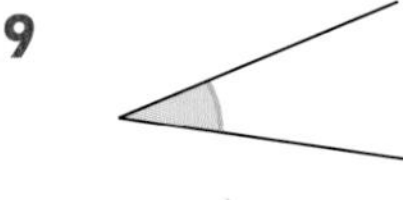

10 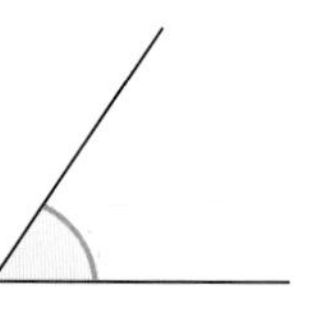

11 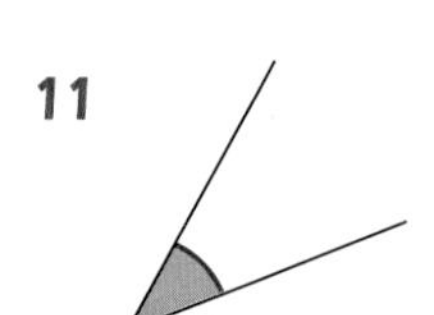

12 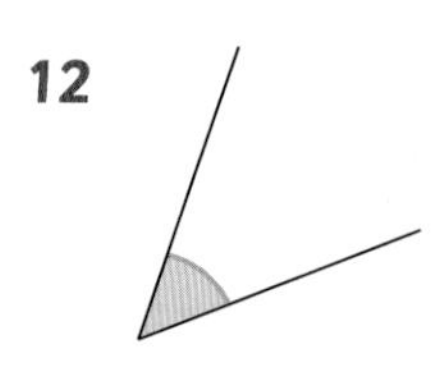

C Write the angle the minute hand on a clock turns through when it moves between each pair of numbers.

13 90 degrees

13 12 and 3
14 7 and 10
15 3 and 9
16 12 and 12
17 6 and 7
18 5 and 7
19 11 and 3
20 10 and 1
21 4 and 11

Challenge

Use a scrap of paper to make an angle measurer.

a Fold the paper to make a right angle.

b Now make half a right angle.

Draw three different shapes. They must each have one right angle and one angle of 45 degrees. Use your folded paper to check the size of the angles.

Do you know that a quarter turn is 90 degrees and a whole turn is 360 degrees?

Review 3

A Use each set of digits to make the largest number possible. Round the number to the nearest 100.

1 5 7 6 2
2 4 3 7 0
3 3 2 9 6
4 4 7 5 6
5 5 9 1 9
6 5 8 0 7

B Answer these.

7 5 + 7 + 4 + 5
8 6 + 6 + 4 + 8
9 80 + 60 + 70
10 86 + 19
11 38 + 69
12 119 + 61

C Copy and complete. Use a written method if you need to.

13 465 + 837
14 523 − 186
15 598 + 362
16 555 + 666
17 731 − 89
18 902 − 478

D Write the area and perimeter of each shape.

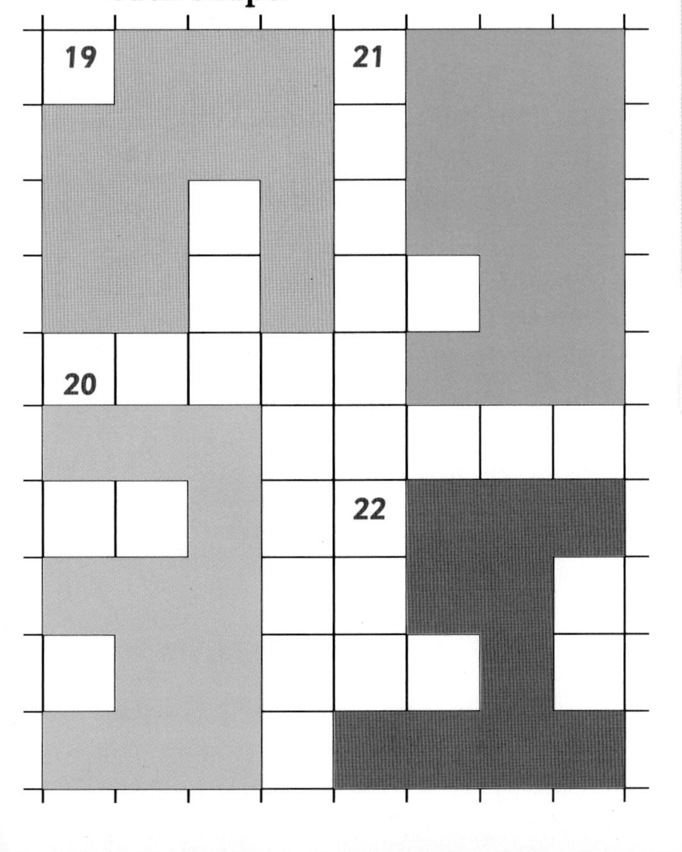

E Solve these word problems.

23 6 chocolate bars cost £1·44. What is the cost of 4 bars?

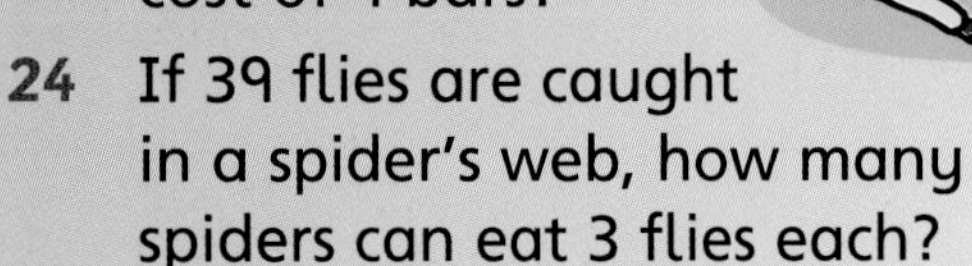

24 If 39 flies are caught in a spider's web, how many spiders can eat 3 flies each?

25 A baker bakes 64 buns on one tray and 47 buns on another. 26 buns burn. How many are left?

26 Declan has £28·40 and his sister has £31·55. Can they afford to buy their mother 2 kittens costing £29·50 each?

F Plot these points on a co-ordinate grid. Join them in order with straight lines. What picture do you find?

27 (1,4) (0,5) (1,5) (0,6) (1,6) (2,7) (3,6) (2,5) (2,4) (5,4) (4,6) (5,5) (5,6) (6,6) (6,5) (7,5) (5,3) (5,2) (4,2) (5,1) (5,0) (4,1) (3,0) (2,0) (3,1) (3,2) (1,2) (1,4)

Properties of numbers and reasoning about numbers

A Write the colour of the counter at each position on the number line.

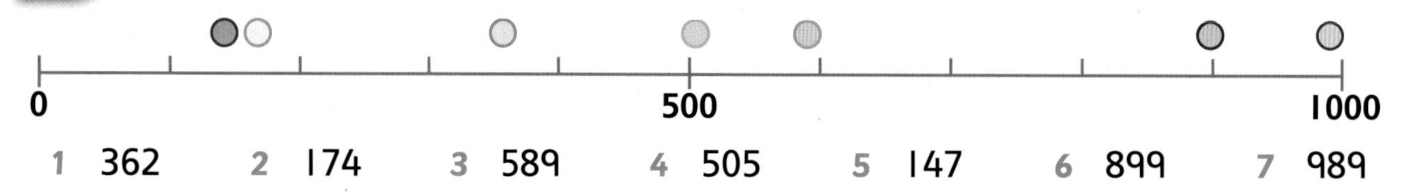

1 362 2 174 3 589 4 505 5 147 6 899 7 989

B Write the numbers missing from each number line in order.

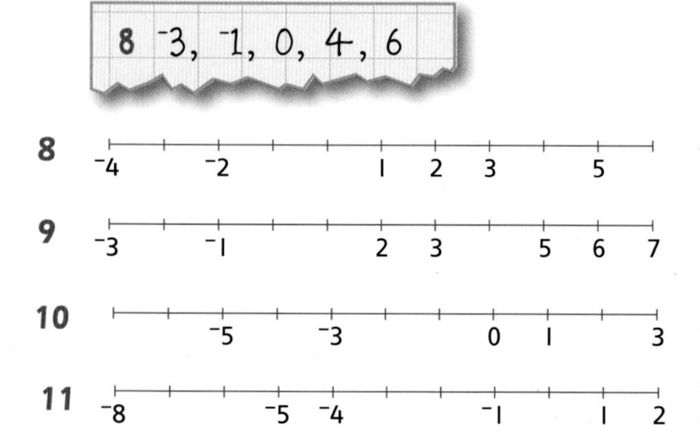

C Write the temperature shown on each thermometer.

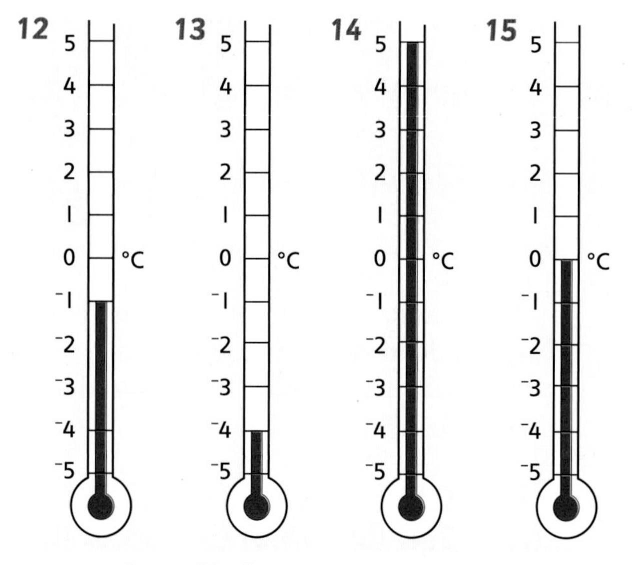

16 Write all the temperatures in order, lowest first.

Challenge

Play with a partner. Each place a counter on the pink 0 °C circle on the thermometer track. Place another counter anywhere on the temperature track. Take turns to roll a dice and move the counter round the temperature track.
If you land on a blue section, the temperature falls by the number of degrees shown on the dice.
If you land on a red section, the temperature rises by the number of degrees shown on the dice.
Move your counter along the thermometer track to match. The first player to reach 10 °C wins, or the first player to reach ⁻10 °C loses.

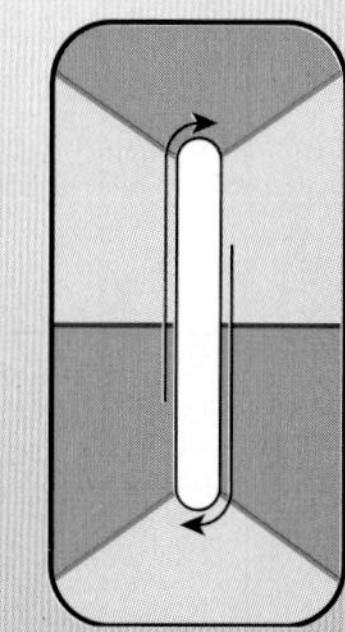

Temperature Track

Thermometer Track

10 9 8 7 6 5 4 3 2 1 0 °C ⁻1 ⁻2 ⁻3 ⁻4 ⁻5 ⁻6 ⁻7 ⁻8 ⁻9 10

Can you find positive and negative numbers on a number line?

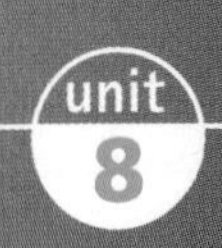

A Explain the rule for each sequence. Write the next three numbers.

1 rule: add 4 each time
11, 15, 19

1. ⁻5, ⁻1, 3, 7, ■, ■, ■
2. 2, 11, 20, 29, ●, ●, ●
3. 16, 10, 4, ⁻2, ▲, ▲, ▲
4. ⁻12, ⁻9, ⁻6, ⁻3, ◖, ◖, ◖
5. ⁻40, ⁻29, ⁻18, ⁻7, ✸, ✸, ✸
6. 24, 18, 12, 6, ⬢, ⬢, ⬢

B Test each general statement with five of your own examples.

7 8 + 9 + 10 = 27,
3 x 9 = 27

7. The sum of three consecutive numbers is three times the middle number.
8. The sum of four odd numbers is an even number.
9. If you draw any square on a hundred square, the sum of each pair of numbers in diagonally opposite corners is the same.

13	14	15	16
23	24	25	26
33	34	35	36
43	44	45	46

10. The number of lines of symmetry in a regular polygon is equal to the number of sides of the polygon.

Challenge

Look at the pattern of posts used in each section of fencing.

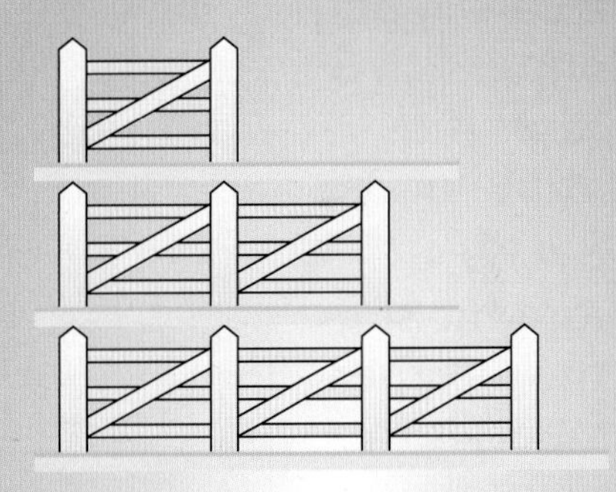

Write how many posts are needed for a fence with:

a 1 section
b 2 sections
c 3 sections.

Write the rule for the fencing.
What general statement can you make?
Test your statement for sections of fencing with more posts.

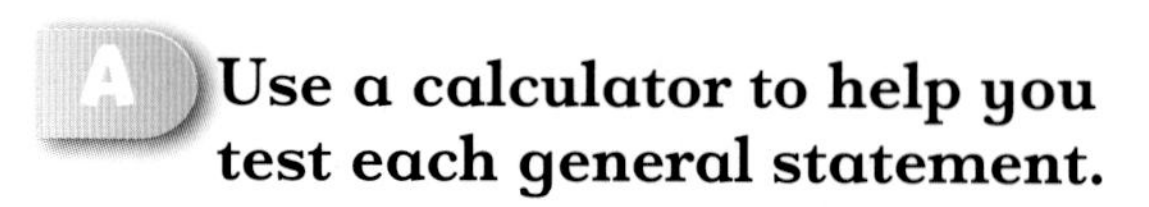

A Use a calculator to help you test each general statement.

1 If you add the digits in any multiple of 3 until you get a single digit, the result will always be 3, 6 or 9.

2 All even numbers are multiples of 2.

3 All multiples of 10 end in '0'.

4 If you add the digits in any multiple of 9 until you get a single digit, the result will always be 9.

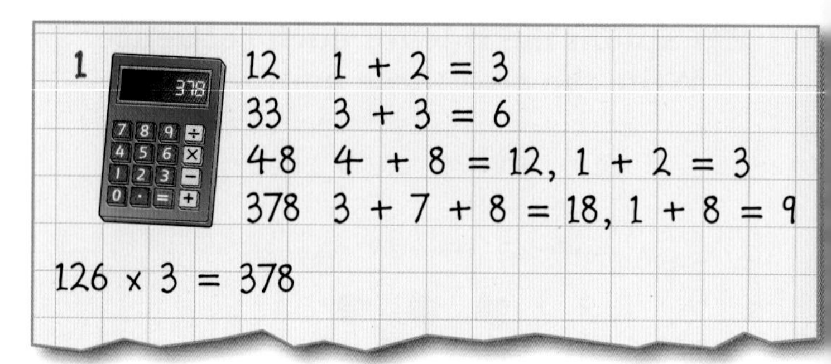

5 The number formed by the last two digits in a multiple of 4 can be divided exactly by 4, e.g. 112, 12 ÷ 4 = 3.

B Use a 100 square to test each general statement.

6 The sum of the numbers in any L-shape with 3 squares is a multiple of 3.

54	
64	65

7 In a 4 by 4 square the sum of the 4 corner numbers is the same as the sum of the 4 centre numbers.

24	25	26	27
34	35	36	37
44	45	46	47
54	55	56	57

8 The sum of the numbers in any L-shape with 9 squares is a multiple of 3.

25				
35				
45				
55				
65	66	67	68	69

9 The sum of the numbers in any 4 by 4 square is a multiple of 4.

Challenge

Work out a general statement for each of these.

a The perimeter of a rectangle when you know its length and breadth.

b The last digit in any multiple of 5.

c The result of adding together three odd numbers.

Can you make and investigate a general statement?

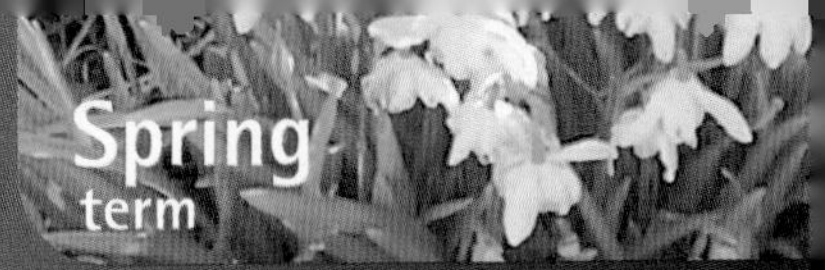

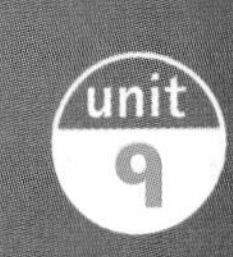

A Write six different number sentences to match each array.

1 5 + 5 + 5 = 15
5 × 3 = 15 3 × 5 = 15
15 ÷ 5 = 3 15 ÷ 3 = 5
3 + 3 + 3 + 3 + 3 = 15

1 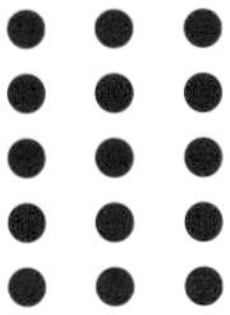

2 

3 

4 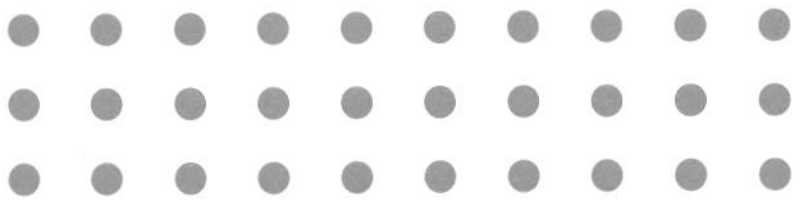

B Answer these.

5 6 × 3
6 5 × 4
7 3 × 5
8 10 × 5
9 8 × 2
10 7 × 4
11 9 × 10
12 9 × 3
13 8 × 4
14 7 × 5
15 9 × 5

C Write two division facts to match each multiplication.

16 7 × 3 = 21
17 8 × 4 = 32
18 7 × 5 = 35
19 4 × 5 = 20
20 9 × 10 = 90

16 21 ÷ 7 = 3
21 ÷ 3 = 7

21 8 × 5 = 40
22 9 × 4 = 36

D Solve these money problems.

23 Share £2 equally among 5 people. How much each?

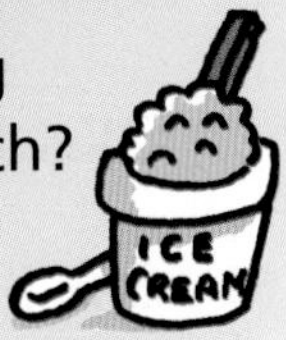

24 Five ice-creams cost £6. What is the cost of 1?

25 How many people can have £4·50 from £23? How much is left over?

26 Jamie has £19. After buying 4 cinema tickets he has £5 left. What is the cost of each ticket?

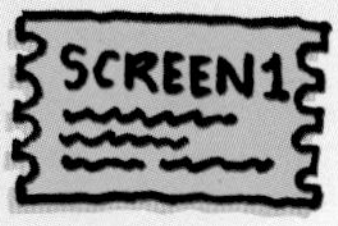

Challenge

Play with a partner. Each place a counter on START.
Take turns to choose an amount from a circle. Roll a dice, move your counter round the track and carry out the instruction. Write down your amount each time.
After 5 turns each, add up your amounts. The winner is the person closest to £100.

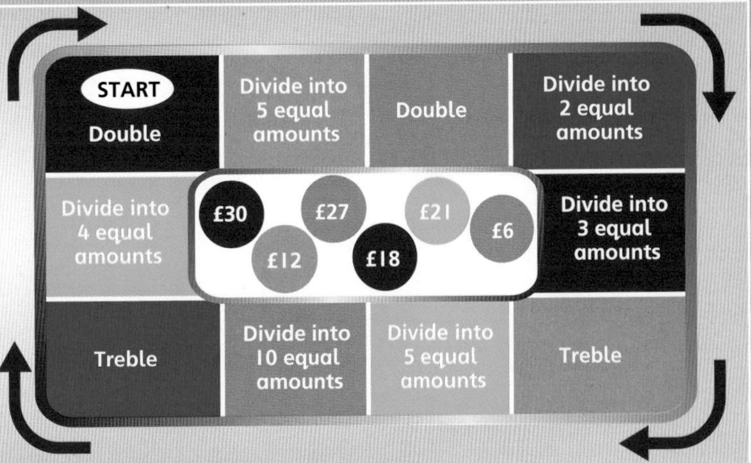

Can you divide a whole number of pounds by 2, 4, 5 or 10?

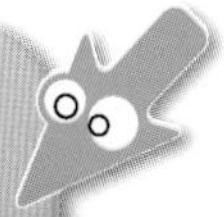

A Answer these.

1 3 times 7
2 4 multiplied by 5
3 half of 23
4 eight fours
5 divide 40 by 5
6 32 shared equally among 4
7 10 multiplied by 10
8 3 times 9
9 divide 28 by 4

B Write the missing numbers.

10 $6 \times 5 =$ ★ $\times 6$
11 $4 \times 6 = 6 \times$ ●
12 $10 \times 5 =$ ▲ $\times 10$
13 $4 \times 7 =$ ● $\times 4$
14 $8 \times 3 = 3 \times$ ✸
15 ⬢ $\times 9 = 9 \times 4$
16 $7 \times$ ◖ $= 3 \times 7$
17 $10 \times$ ■ $= 6 \times 10$
18 ★ $\times 5 = 5 \times 8$

C Copy and complete.

19 $6 \times 2 \times 3$
20 $4 \times 7 \times 2$
21 $2 \times 10 \times 6$
22 $3 \times 5 \times 4$
23 $10 \times 6 \times 2$
24 $5 \times 7 \times 4$
25 $2 \times 9 \times 5$

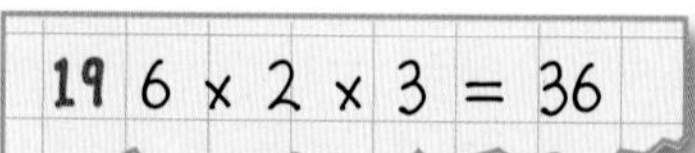

D Write the missing number for each.

26 $4 \times 15 = 4 \times$ ★ $\times 5$
27 $8 \times 20 = 8 \times 2 \times$ ●
28 $5 \times 12 = 5 \times$ ⬢ $\times 6$
29 $9 \times 12 = 9 \times 4 \times$ ▲
30 $7 \times 20 = 7 \times$ ● $\times 10$

Challenge

Answer the questions and follow the trail. What do you find at the end?

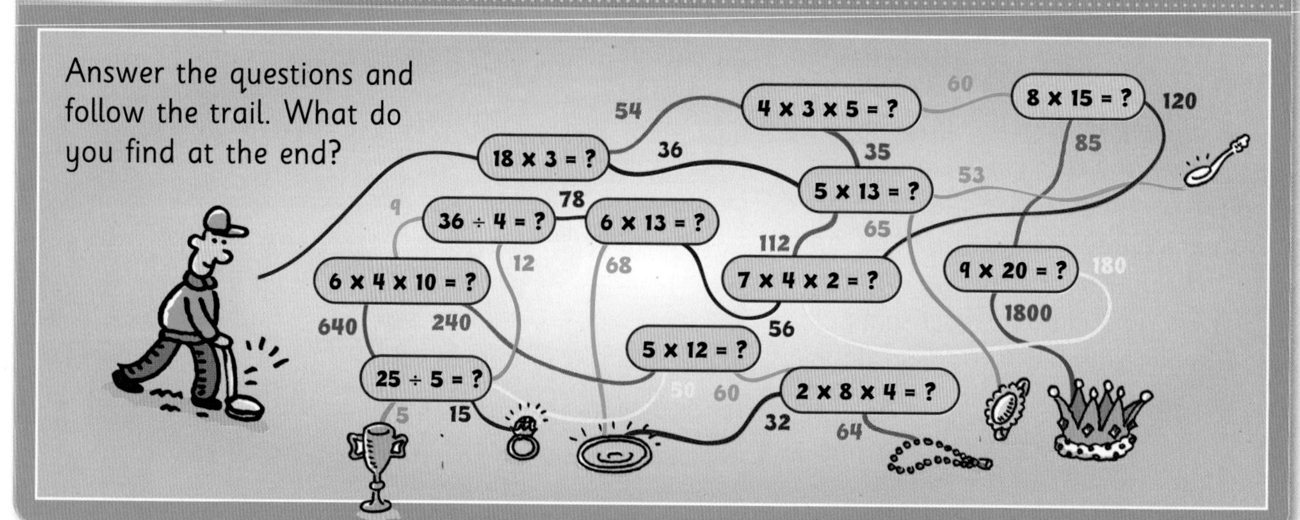

Do you know that multiplications can be done in any order?

A Write the missing numbers.

1 (5 × 3) + (5 × 2) = 5 × ★
2 7 × 6 = (7 × 4) + (7 × ▲)
3 (8 × 2) + (8 × 5) = 8 × ■
4 9 × 8 = (9 × 4) + (9 × ⬢)
5 (6 × 5) + (6 × ●) = 6 × 9
6 (9 × 2) + (✸ × 3) = 9 × 5

B Write the missing numbers.

7 2, 60

7 5 × 12 = (5 × 10) + (5 × ●) = ⬢
8 6 × 13 = (6 × ●) + (6 × 3) = ■
9 7 × 8 = (7 × 2) + (7 × ■) = ★
10 8 × 9 = (8 × 10) − (8 × ●) = ■
11 9 × 5 = (✸ × 3) + (9 × 2) = ●
12 11 × 9 = (11 × 10) − (11 × ✸) = ✸

C Draw a grid to answer these.

13 7 × 15
14 6 × 12
15 8 × 14
16 9 × 13
17 20 × 18
18 7 × 13
19 4 × 24

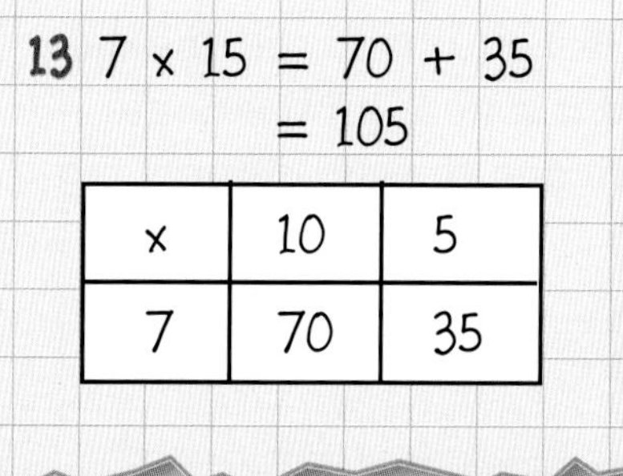
13 7 × 15 = 70 + 35
= 105

×	10	5
7	70	35

D Copy and complete.

20 36 × 3 = (30 × 3) + (6 × 3) = ■
21 42 × 4 = (40 × 4) + (2 × 4) = ●
22 57 × 3 = ●
23 48 × 2 = ✸
24 59 × 4 = ■
25 63 × 5 = ▲
26 88 × 3 = ■
27 79 × 5 = ✸
28 67 × 4 = ●
29 36 × 5 = ✸

Challenge

Cover each wrong answer on the grid with a counter. What number do the counters make?

4 × 12 = 52	14 × 11 = 131	7 × 8 = 65
8 × 8 = 56	25 × 2 = 50	11 × 11 = 111
18 × 9 = 189	5 × 6 × 3 = 80	7 × 13 = 81
17 × 5 = 85	9 × 7 = 63	8 × 15 = 100
5 × 3 × 2 = 30	5 × 6 × 2 = 60	3 × 24 = 75
9 × 12 = 108	6 × 4 × 3 = 72	20 × 16 = 160

Can you use partitioning to multiply a 2-digit number by a single-digit number?

Right angle, wrong angle

You need:
- a partner
- a dice
- two sets of 10 counters

START (throw again)

right angle 6 × 2 × 3

½ right angle 12 × 2

less than 45 degrees 5 × 2 × 4

½ right angle 8 × 5

greater than a right angle (6 × 10) − (6 × 4)

less than 45 degrees (6 × 5) + (6 × 5)

right angle 2 × 10 × 2

45 degrees 80 ÷ 10

greater than a right angle 32 ÷ 4

less than 45 degrees (3 × 10) + (3 × 2)

greater than 90 degrees 10 × 2 × 3

90 degrees 12 × 5

less than 45 degrees 40 ÷ 5

right angle 48 ÷ 2

greater than 90 degrees 9 × 4

½ right angle 2 × 5 × 6

greater than a right angle 8 × 3

36	40	24	60
8	36	24	40
8	24	40	8
60	60	36	60

Each place a counter on START. Take turns to roll the dice and move around the track. Find a square on the angle grid that matches both the angle and the calculation, e.g.

right angle, 6 × 2 × 3 matches 36,

If you can find a matching square, place a counter it.
Now it is your partner's turn.
The first player to make a line of three counters on the grid is the winner.

A Answer these.

1. 6 × 4
2. 7 × 5
3. 40 ÷ 5
4. double 16
5. $\frac{1}{2}$ of 440
6. 44 ÷ 2
7. 34 × 2
8. 14 × ● = 28
9. ● ÷ 2 = 26
10. 5 × 3 × 2
11. 6 × 3 × 2
12. 5 × 7 × 4
13. 20 × 4

B Copy and complete the tables. Write approximate answers for each, then use a grid method to find the exact answers. Find the difference between the approximate and exact answers.

		estimate	answer	difference
14	62 × 8	480	496	16
15	68 × 5			
16	43 × 8			
17	74 × 3			
18	93 × 4			
19	34 × 7			
20	51 × 9			

		estimate	answer	difference
21	76 ÷ 4			
22	87 ÷ 3			
23	115 ÷ 5			
24	96 ÷ 6			
25	124 ÷ 4			
26	104 ÷ 8			
27	98 ÷ 7			

Challenge

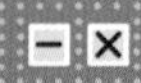

Play with a partner. Place a counter on START at the snake's tail. Take turns to roll the dice and move the counter along the snake.
When you land on a question, each write down an approximate answer. Work out the correct answer together. The player with the closer estimate scores 1 point. Continue until one player scores 5 points to win. If you land on a question you have already answered, roll again.

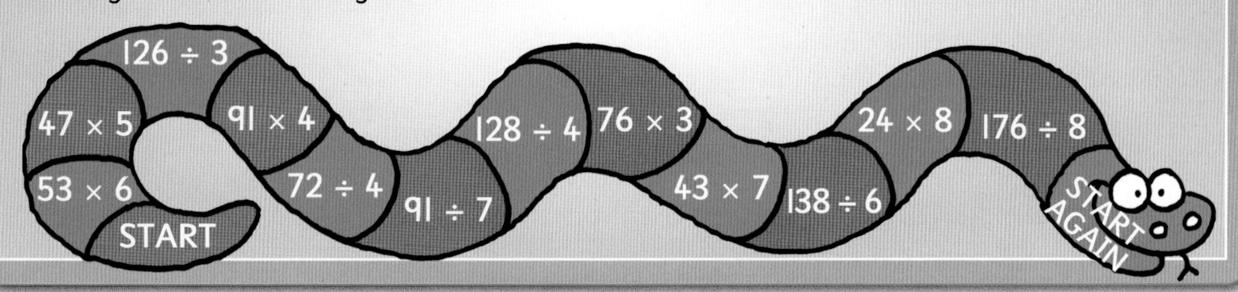

Can you make good approximations for multiplication and division problems?

A Solve these word problems. Give an approximate answer first. Explain what the remainder means for each.

1 estimate: 10 weeks
exact answer: 9 r 1
9 weeks and 1 day left over

1 The spring term lasts for 64 days. How many weeks is that?

2 How many pounds each if 33 pound coins are shared among 4 children?

3 How many 20 cm ribbons can be cut from a 365 cm roll?

4 Minibuses carry 15 people. How many buses are needed to carry 100 people?

5 How many horses can be fitted with new shoes if a blacksmith has 69 horseshoes?

B Solve these word problems.

6 A spider has 8 legs. How many legs on 14 spiders?

7 Jesminder has 3 boxes of marbles. There are 47 in one box and 48 in each of the other boxes. How many marbles does she have altogether?

8 Cakes are packed in boxes of 6. How many boxes do 114 cakes fill?

9 In a Greek restaurant, 38 plates are broken every evening. How many plates are broken in a week?

10 A greengrocer has 60 bags of onions to sell. 6 people buy 3 bags each. How many bags of onions are left?

Challenge

Work out the remainder for each division. Use the Remainder Code Box to find the secret password.

a $28 \div 5$ **b** $95 \div 3$ **c** $86 \div 9$ **d** $100 \div 3$ **e** $111 \div 15$ **f** $509 \div 10$ **g** $95 \div 7$ **h** $53 \div 8$

Make up your own code to give a different password.

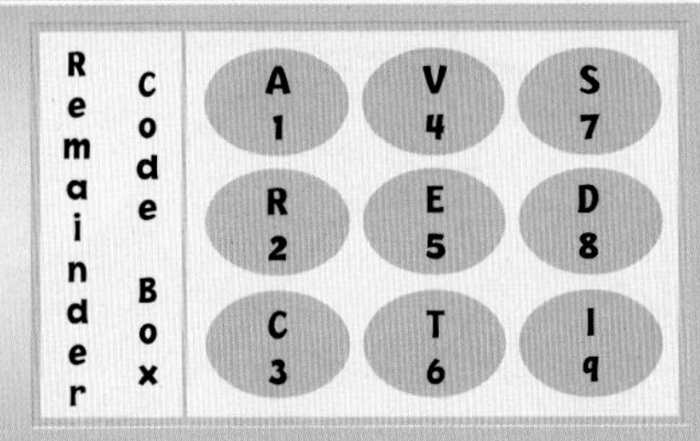

Remainder Code Box

A 1	V 4	S 7
R 2	E 5	D 8
C 3	T 6	I 9

Can you choose the appropriate operation when solving a problem?

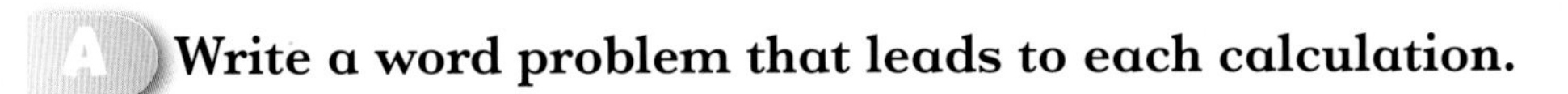

A Write a word problem that leads to each calculation.

1 $13 \times 6 = 78$
2 $64 + 36 = 100$
3 $132 - 90 = 42$
4 $16 \times 7 = 112$
5 $320 \div 8 = 40$

1 Doughnuts are packed in boxes of 6. How many doughnuts in 13 boxes?

B Write a story for each time line.

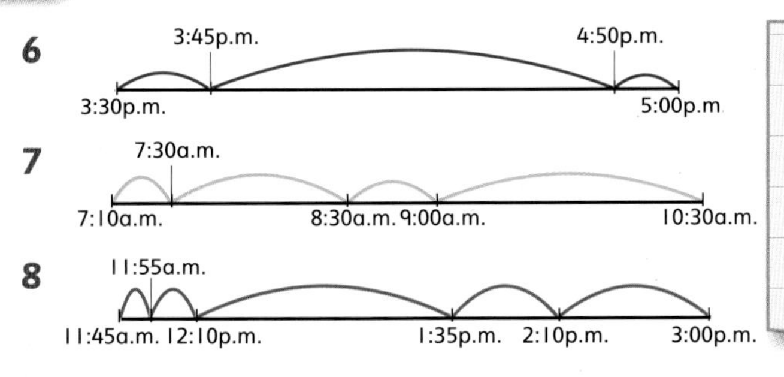

6 Abi left school at 3:30 p.m. and walked for 15 minutes to reach the park. She played with her friends in the park for 1 hour and 5 minutes, then took 10 minutes to walk home.

C Solve these time problems.

9 Stan watches T.V. from 4:43 p.m. until 5:17 p.m. For how long is he watching T.V.?

10 A school concert starts at 1:45 p.m. and finishes at 3:08 p.m. How long does the concert last?

11 Nico arrives home at 4:40 p.m. He starts his homework 35 minutes later and works for 1 hour and 25 minutes. What time does he finish his homework?

12 Buses leave at 17 minutes past and 47 minutes past each hour. Jack arrives at the bus stop at 6:32 p.m. His journey takes 37 minutes. What time does he get off the bus?

13 Trains leave every 42 minutes. The first train leaves at 7:26 a.m. What time do the next three trains leave?

Challenge

Follow the blue line to see how Bluewater becomes 48. Work out numbers for the other stations.

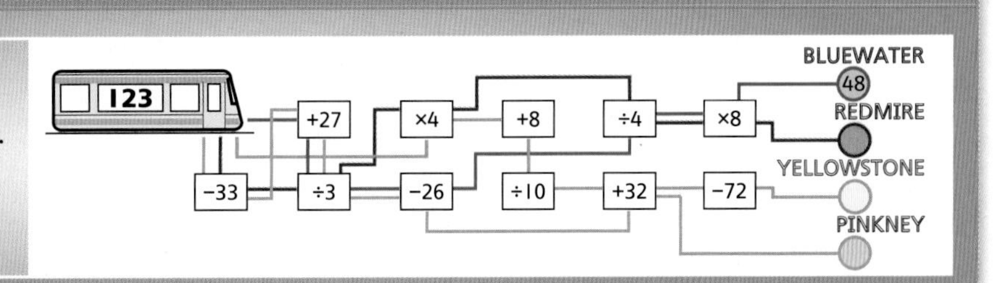

A Answer these.

1. $\frac{1}{3}$ of £24
2. $\frac{1}{2}$ of £80
3. $\frac{1}{5}$ of £35
4. $\frac{3}{4}$ of 48 cm
5. $\frac{3}{5}$ of 30 buns
6. $\frac{5}{8}$ of 40 goals
7. $\frac{7}{10}$ of 120 days
8. $\frac{2}{3}$ of 33 cars
9. $\frac{2}{5}$ of 60 buses
10. $\frac{1}{4}$ of 320 sheep
11. $\frac{7}{8}$ of 64 counters

B Write what fraction of each shape is green and what fraction is yellow.

12 $\frac{3}{8}$ green, $\frac{5}{8}$ yellow

12

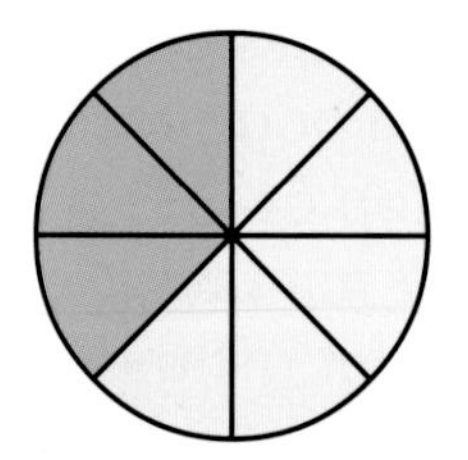

15

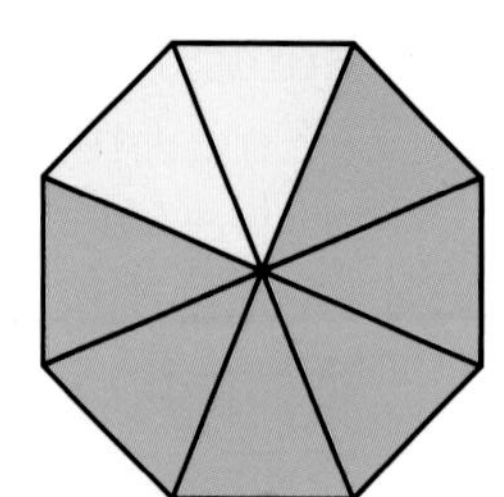

13

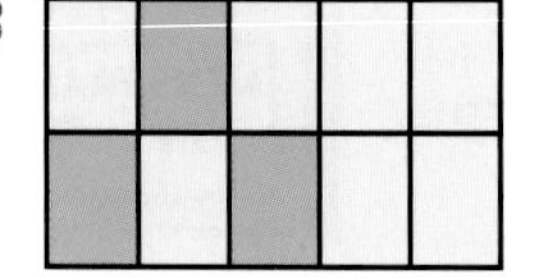

16

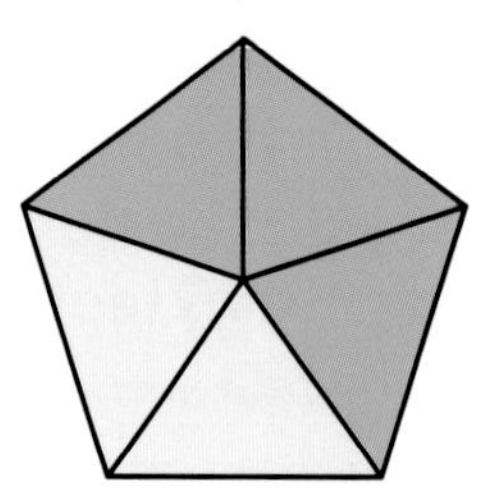

14

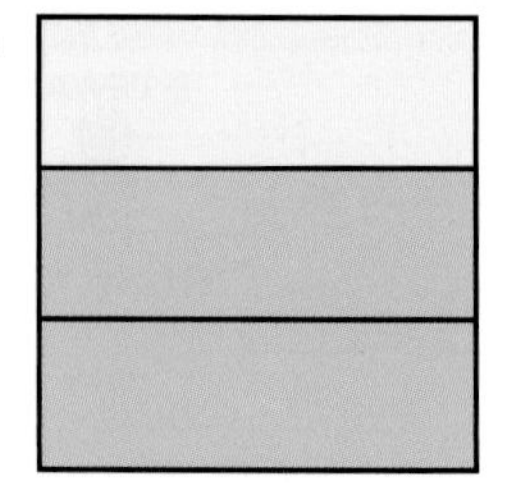

17

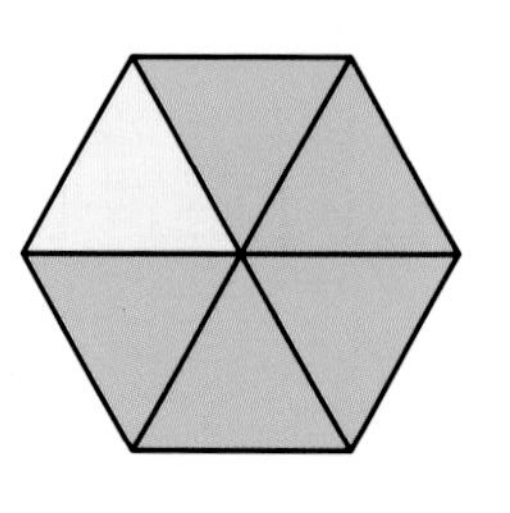

C Write a fraction statement for each shape in section B.

18 $\frac{3}{8} + \frac{5}{8} = \frac{8}{8} = 1$ whole

18 circle
19 square
20 rectangle
21 pentagon
22 hexagon

D Copy and complete.

23 $\frac{1}{6} + \frac{\square}{6} = 1$ whole
24 $\frac{3}{4} + \frac{\square}{4} = 1$ whole
25 $\frac{\square}{5} + \frac{2}{5} = 1$ whole
26 $\frac{7}{10} + \frac{3}{\square} = 1$ whole

E Copy and complete.

27 $\frac{1}{2} = \frac{4}{\square}$
28 $\frac{2}{5} = \frac{\square}{10}$
29 $\frac{2}{3} = \frac{\square}{6}$
30 $\frac{4}{8} = \frac{\square}{2}$
31 $\frac{4}{6} = \frac{2}{\square}$
32 $\frac{5}{10} = \frac{1}{\square}$
33 $\frac{3}{4} = \frac{\square}{8}$

F Write the missing signs. Use < or >.

34 $\frac{1}{2} \square \frac{1}{4}$
35 $\frac{1}{4} \square \frac{1}{3}$
36 $\frac{1}{5} \square \frac{1}{2}$
37 $\frac{1}{8} \square \frac{1}{10}$
38 $\frac{1}{5} \square \frac{1}{6}$
39 $\frac{1}{10} \square \frac{1}{4}$

Challenge

Write 'true' or 'false' for each pair of fractions. Explain how you worked out your answer.

a $\frac{2}{5} < \frac{3}{4}$
b $\frac{5}{8} < \frac{3}{10}$
c $\frac{3}{5} > \frac{1}{3}$
d $\frac{5}{6} < \frac{3}{4}$

A Write a fraction and a decimal fraction to show how much of each shape is coloured.

1 $\frac{7}{10}$ = 0·7

1 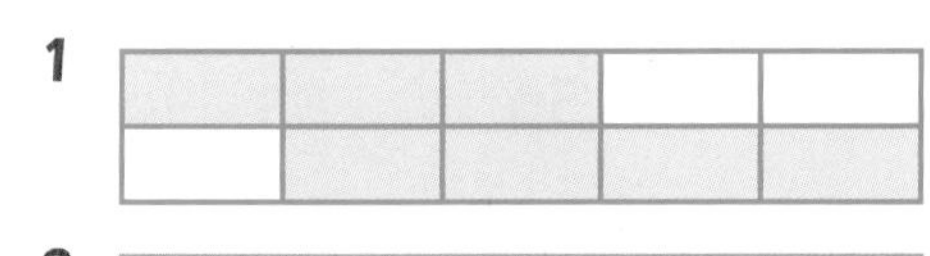

2 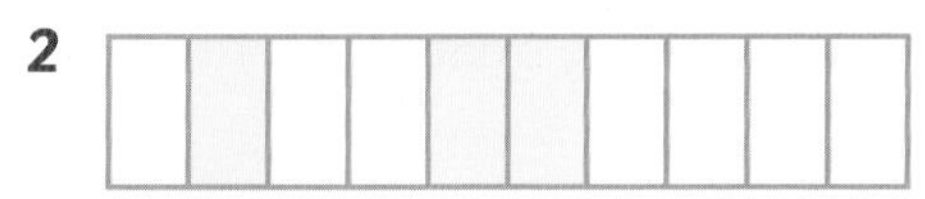

3 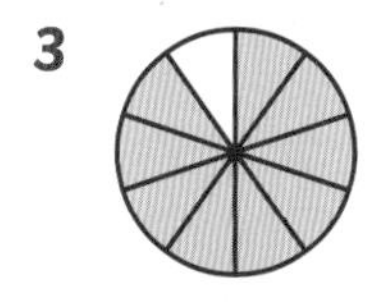

4 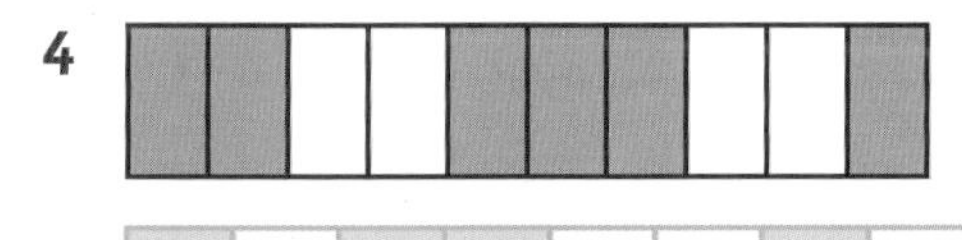

5 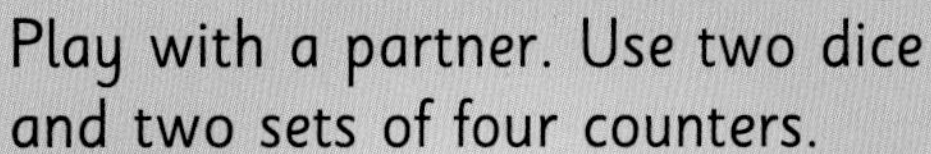

B Write each fraction as a decimal fraction.

6 $\frac{3}{10}$ 8 $\frac{9}{10}$ 10 $\frac{5}{10}$

7 $\frac{6}{10}$ 9 $\frac{1}{10}$ 11 $\frac{8}{10}$

C Write each decimal fraction as a fraction.

12 0·7 14 0·6 16 0·1

13 0·3 15 0·9 17 0·5

D Write how many tenths.

18 3 whole ones
19 2 whole ones
20 5 whole ones
21 7 whole ones
22 6 whole ones
23 8 whole ones
24 10 whole ones

E Write as decimal fractions.

25 16 tenths
26 14 tenths
27 18 tenths
28 23 tenths
29 44 tenths

25 1·6

F Write each set of decimal fractions in order, smallest first.

30 0·5 0·3 2·6 3·2 0·9
31 0·9 1·2 2·1 1·9 0·2
32 1·6 2·4 0·7 6·1 0·6
33 3·2 2·3 0·3 0·2 2·2
34 6·8 8·6 5·8 5·6 8·5

Challenge

Play with a partner. Use two dice and two sets of four counters.
Take turns to roll the dice and use the two numbers to make a fraction.

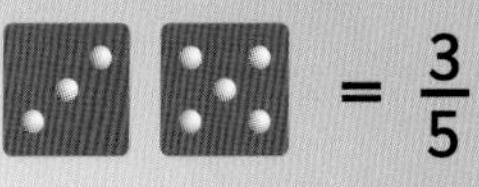 = $\frac{3}{5}$

If you can find a square on the grid with the same value, cover it with a counter.
The first player to have all four counters on the grid wins.

0·5	0·4	$\frac{2}{6}$	0·8
one half	one whole	0·6	$\frac{6}{8}$
0·2	$\frac{2}{3}$	$\frac{4}{10}$	$\frac{4}{6}$
$\frac{5}{10}$	$\frac{1}{2}$	one whole	0·4

Can you write a fraction equivalent to a decimal fraction?

A Write a fraction and a decimal fraction to show how much of each shape is coloured.

1 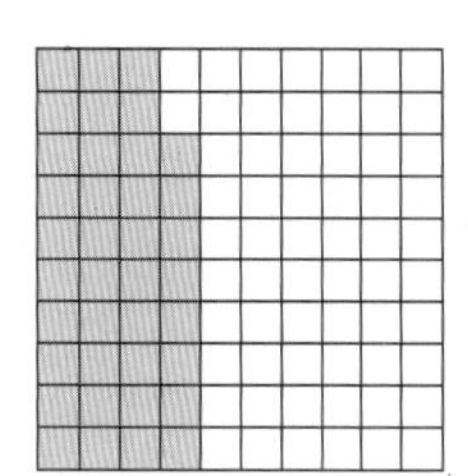

1 $\frac{38}{100}$ = 0·38

2 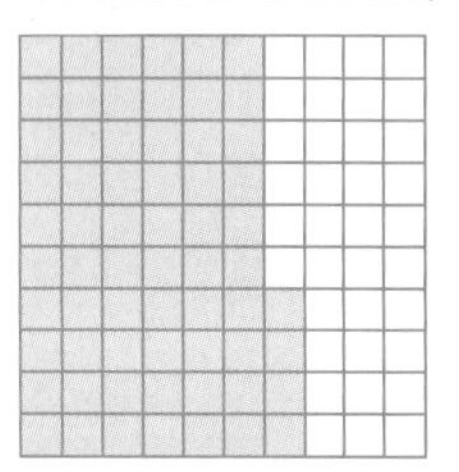

4 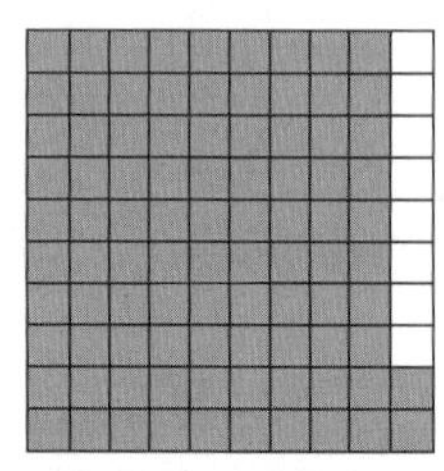

3 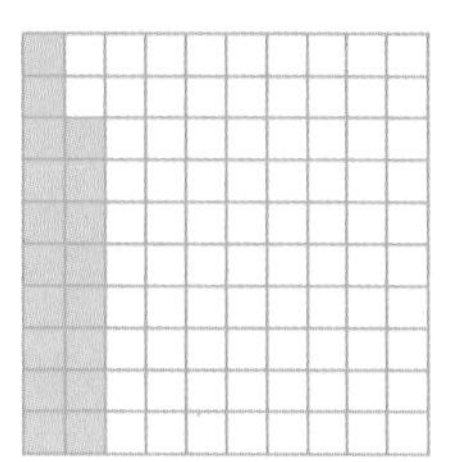

5 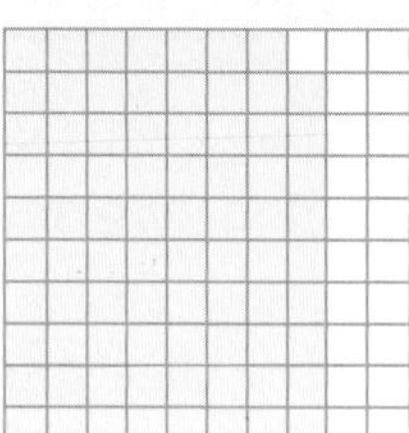

B Write as decimal fractions.

6 $\frac{29}{100}$ 7 $\frac{36}{100}$ 8 $\frac{45}{100}$ 9 $\frac{93}{100}$ 10 $\frac{78}{100}$ 11 $\frac{61}{100}$ 12 $\frac{16}{100}$

C Copy and complete.

13 $\frac{40}{100} = \frac{●}{10}$

14 $\frac{30}{100} = \frac{✷}{10}$

15 $\frac{60}{100} = \frac{●}{10}$

16 $\frac{70}{100} = \frac{✹}{10}$

17 $\frac{20}{100} = \frac{2}{●}$

18 $\frac{50}{■} = \frac{5}{10}$

19 $\frac{90}{100} = \frac{9}{●}$

D Write as tenths and hundredths.

20 $\frac{61}{100}$ 21 $\frac{54}{100}$ 22 $\frac{27}{100}$ 23 $\frac{96}{100}$ 24 $\frac{33}{100}$ 25 $\frac{79}{100}$

20 6 tenths
1 hundredth

E Write each amount in pounds.

26 £1·37

26 137p 27 128p 28 284p 29 28p 30 737p 31 15p 32 3p 33 9p

F Write in order, smallest first.

34	1·25	1·68	0·37	0·92	1·29
35	0·68	1·32	0·86	2·31	0·33
36	3·62	0·62	6·02	6·32	2·36
37	5·12	5·21	5·32	5·23	5·02
38	6·62	6·63	6·66	6·06	6·26
39	3·95	3·59	4·59	4·95	3·49
40	0·6	0·82	0·68	0·86	0·8
41	0·7	1·32	0·4	1·23	0·47
42	1·12	2·11	1·11	1·21	0·12

Challenge

On a 10 by 10 grid, create an interesting picture by colouring in squares. Follow these rules:

a 0·3 of the grid must be red
b 0·2 of the grid must be blue
c 0·2 of the grid must be yellow
d 0·2 of the grid must be black
e 0·1 of the grid must be white.

Can you write tenths and hundredths as decimal fractions?

You need:

- a partner
- a dice
- a counter

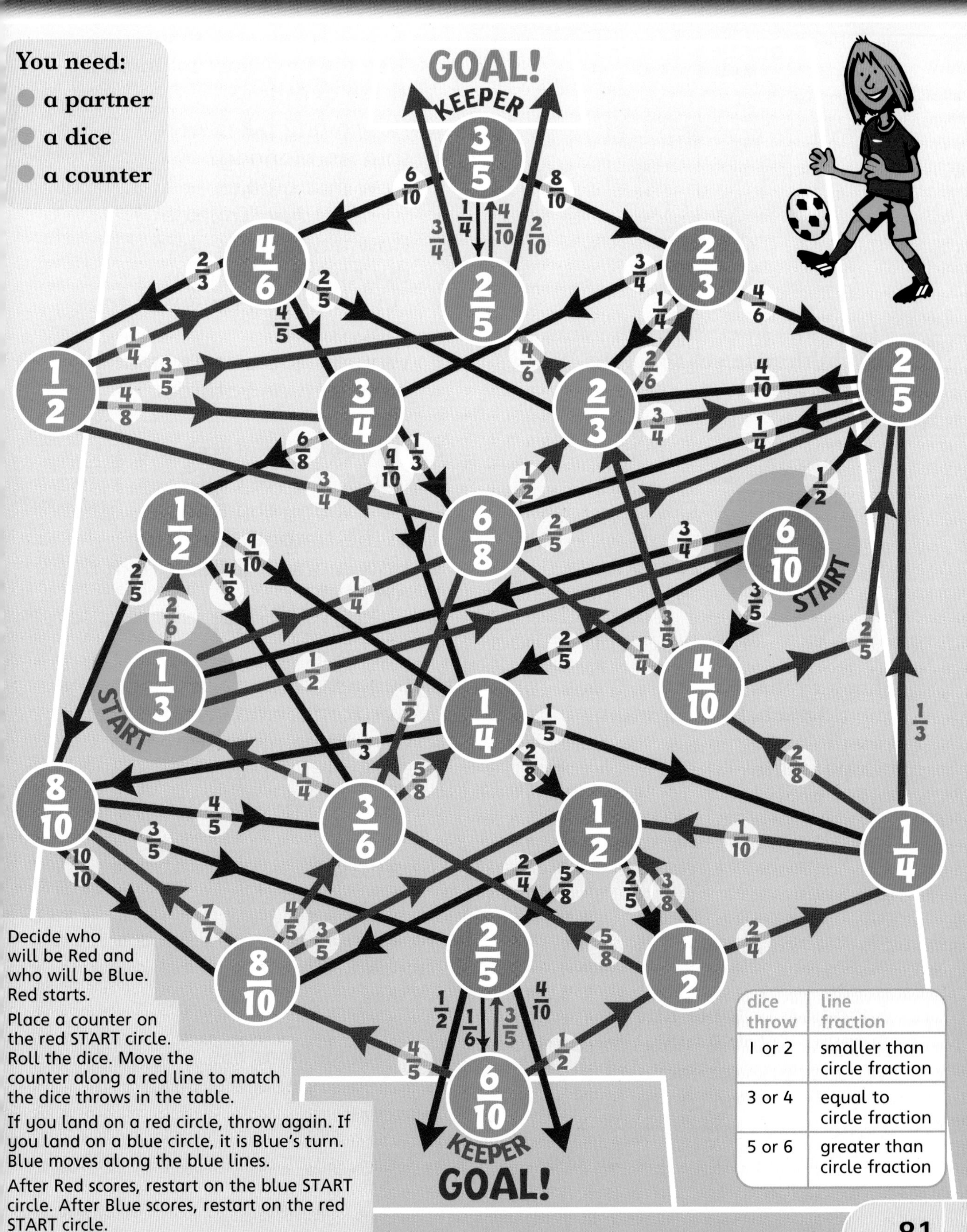

Decide who will be Red and who will be Blue. Red starts.

Place a counter on the red START circle. Roll the dice. Move the counter along a red line to match the dice throws in the table.

If you land on a red circle, throw again. If you land on a blue circle, it is Blue's turn. Blue moves along the blue lines.

After Red scores, restart on the blue START circle. After Blue scores, restart on the red START circle.

The first player to score three goals wins.

dice throw	line fraction
1 or 2	smaller than circle fraction
3 or 4	equal to circle fraction
5 or 6	greater than circle fraction

A

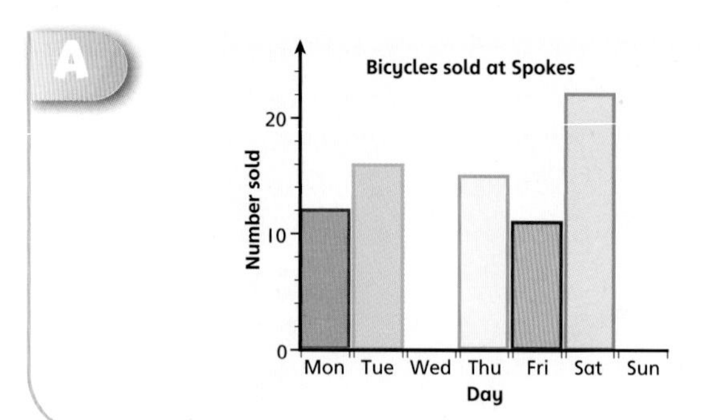

Use the bar chart to answer these questions.

1 12

1 How many bikes were sold on Monday?
2 How many bikes were sold on Thursday?
3 How many bikes were sold during the week?
4 On which two days was the shop closed?
5 Why do you think so many bikes were sold on Saturday?

B

This bar chart shows the number of children in school clubs.

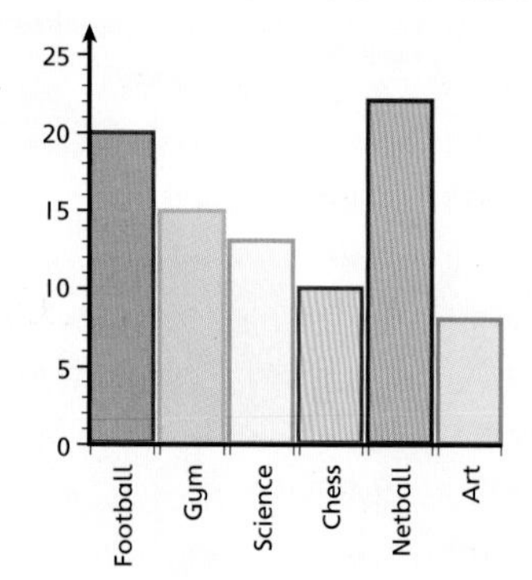

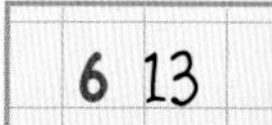

6 How many children are in the Science Club?
7 How many children are in the Netball Club?
8 How many children are in the Art Club?
9 Suggest a suitable title for the bar chart.
10 Suggest suitable labels for the horizontal and vertical axes.
11 Give three more statements about information you can gather from the bar chart.

C

Look at this bar chart. It has no title, labels, or marking on the axes. Copy it into your book.

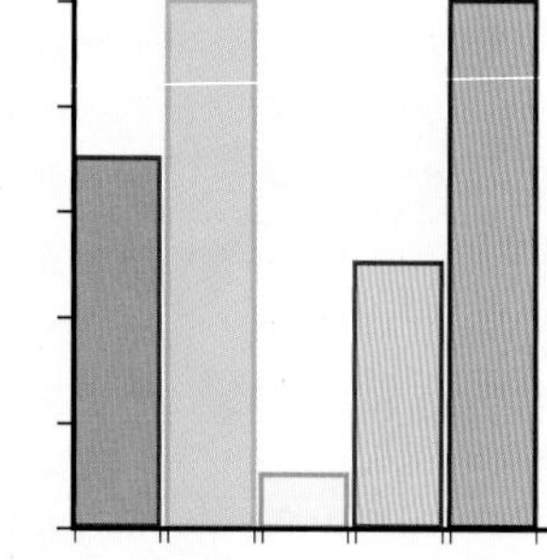

12 Suggest a suitable title.
13 Suggest suitable labels for the axes.
14 Write details on each axis so that the information can be read easily.
15 Write three statements giving information from your bar chart.

Challenge

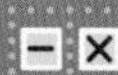

Make two copies of the bar chart in section C. Complete the bar charts to show information about:

a the weather
b a school sports day.

Write three statements about each bar chart.

Can you answer questions using information given on bar charts?

A Use the Venn diagram to answer these questions.

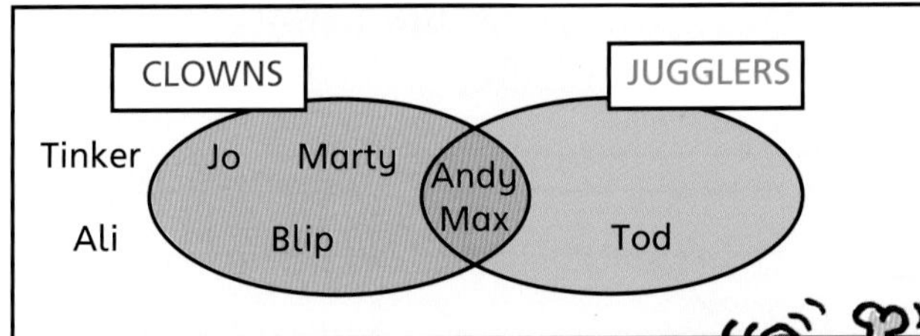

1. How many jugglers are there?
2. How many clowns are there?
3. How many people are both clowns and jugglers?
4. How many people are neither clowns nor jugglers?
5. How many clowns do not juggle?

B Copy and complete each Venn diagram by writing in the set of numbers.

6 numbers 1 to 30

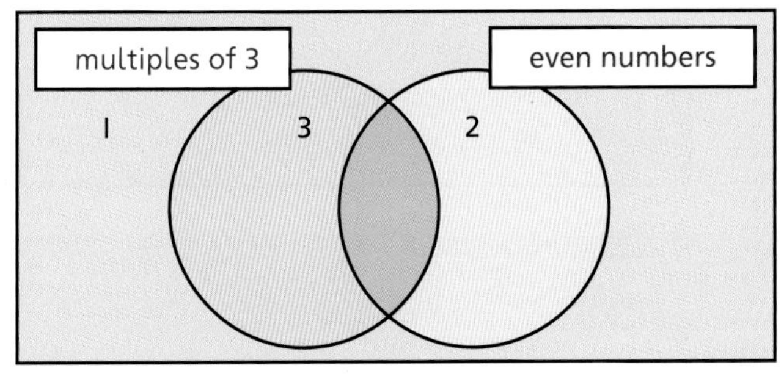

7 numbers 50 to 100

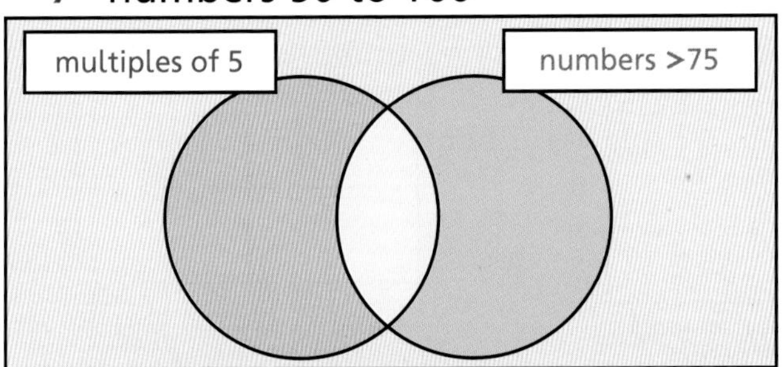

C Copy and complete the Carroll diagrams with each set of numbers.

8 numbers 20 to 40

	odd	not odd
numbers greater than 30		
numbers not greater than 30	21	20

9 numbers 31 to 69

	multiple of 3	not a multiple of 3
numbers with at least 1 even digit		
numbers that do not have any even digits		

Challenge

Play with a partner. Take turns to roll two dice and add the numbers together. Work out where the answer fits on the Venn diagram.

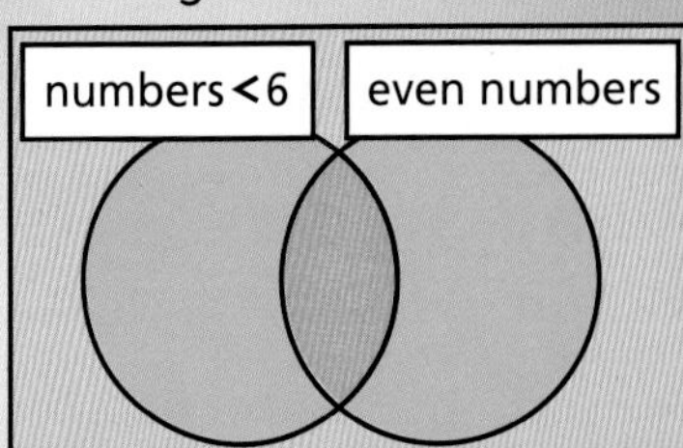

Score 5 points for a number in the orange section. Score 2 points for a number in the green sections. Lose 1 point for a number in the pink section. The first player to score 25 points wins.

Design a game using different labels for the Venn diagram.

A

Copy the Carroll diagram. Write letters in the sections to show where each shape fits.

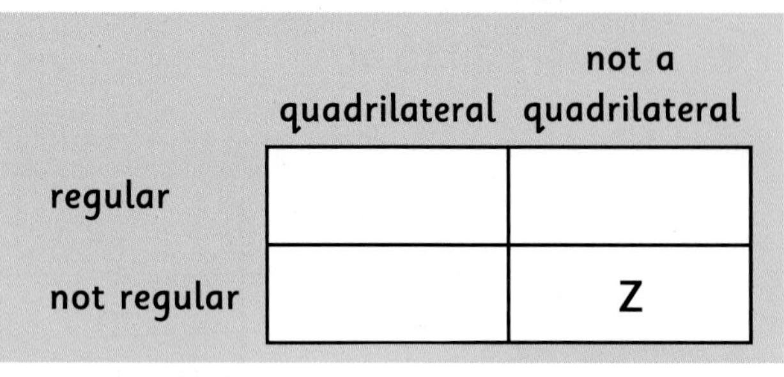

	quadrilateral	not a quadrilateral
regular		
not regular		Z

1. Z
2. Y
3. X
4. W
5. V
6. U
7. T
8. S
9. R

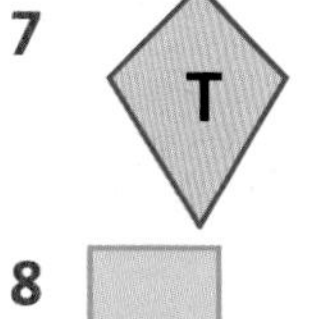

B

These Venn diagrams each contain the numbers 1 to 20. Work out possible labels for each circle.

10

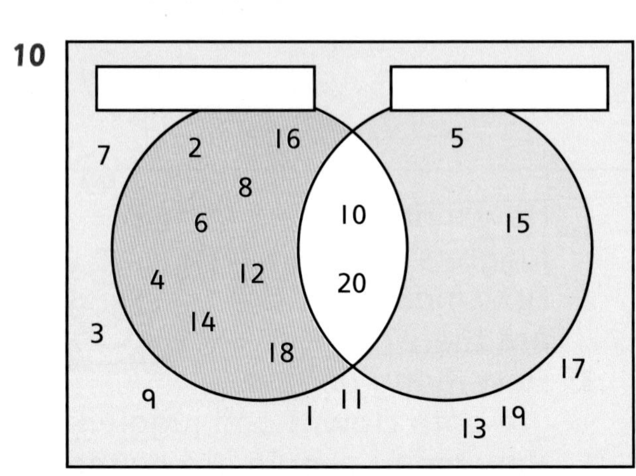

11

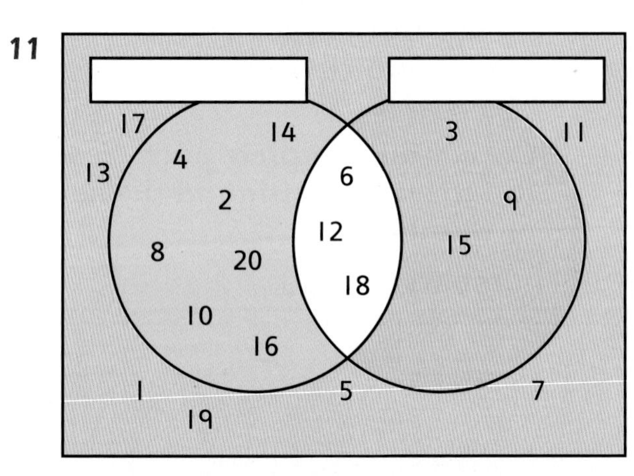

Challenge

Copy the Venn and Carroll diagrams. Enter the initials of every child in your class on each diagram. If any section in either diagram is empty, explain why.

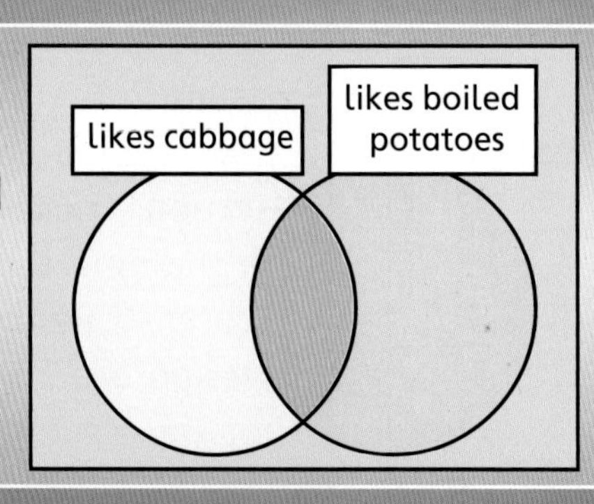

	older than you	not older than you
is a boy		
is not a boy		

Can you gather information from data shown in Venn and Carroll diagrams?

Review 4

A Write the numbers missing from each sequence.

1 2, 8, 14, ▲, ▲, 32, ▲

2 ⁻16, ⁻9, ●, ●, 12, 19

3 ⁻21, ⁻15, ✸, ✸, ✸, 9, 15

4 11, 3, ★, ★, ⁻21, ⁻29

5 12, 6, ■, ■, ⁻12, ⁻18, ■

B Write the missing numbers.

6 $3 \times 8 = ✸ \times 3$

7 $4 \times 2 \times 6 =$ ★

8 $5 \times 3 \times 5 =$ ●

9 $8 \times 30 =$ ■

10 $7 \times 13 =$ ■

11 $8 \times 14 =$ ●

12 $7 \times 8 = (7 \times 6) + (7 \times$ ⬢)

13 $9 \times 9 = (9 \times 10) - (9 \times$ ●)

14 $46 \times 3 =$ ●

15 $68 \times 4 =$ ✸

C Solve these problems.

16 Light bulbs are packed in boxes of 4. How many boxes can be filled with 108 bulbs?

17 A clown bursts 26 balloons in each performance. How many balloons are burst in 12 performances?

18 A train journey starts at 6:42 a.m. and finishes at 8:17 a.m. How long does the journey take?

19 Arni watches T.V. from 5:25 p.m. until 6:08 p.m. and from 6:42 p.m. until 7:36 p.m. How long does he watch T.V. altogether?

D Copy and complete.

20 $\frac{3}{5} = \frac{●}{10}$

21 $\frac{4}{⬢} = \frac{1}{2}$

22 $\frac{3}{5} + \frac{●}{5} =$ 1 whole

23 $\frac{▲}{10} = \frac{1}{5}$

24 $\frac{3}{6} = \frac{●}{2}$

25 $\frac{8}{✸} = \frac{4}{5}$

E Write the missing sign. Use < or >.

26 $\frac{1}{2}$ ● $\frac{1}{3}$

27 $\frac{3}{10}$ ★ $\frac{1}{2}$

28 $\frac{3}{4}$ ✸ $\frac{3}{5}$

29 $\frac{2}{5}$ ● $\frac{5}{8}$

F Write as fractions.

30 0·6 31 0·2 32 0·9 33 0·5

G Write as decimal fractions.

34 $\frac{7}{10}$ 35 $\frac{9}{10}$ 36 $\frac{23}{100}$ 37 $\frac{77}{100}$

H Copy and complete this Venn diagram. Enter numbers 21 to 54.

38

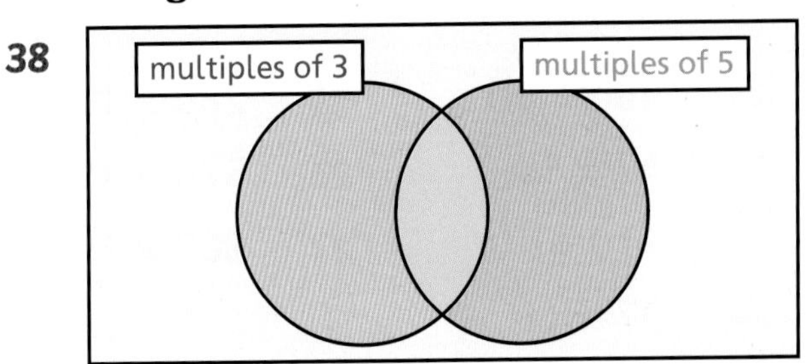

A Write each set of numbers in order, smallest first.

1	206	360	260	602	306
2	412	214	341	241	212
3	51	60	106	56	65
4	149	109	401	104	419
5	1000	100	101	111	110

B Round to the nearest 100.

6 848 8 827 10 390 12 950
7 119 9 656 11 450

C Find which balloon each arrow bursts.

13 62 × 100 = 6200, yellow balloon

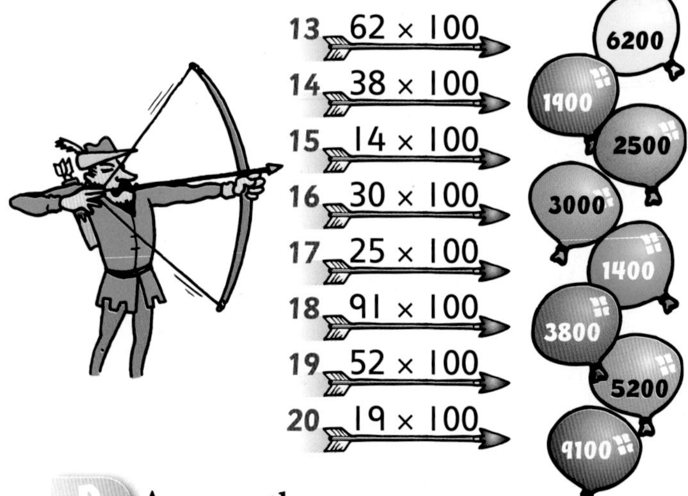

D Answer these.

21 460 ÷ 10
22 42 × 10
23 710 ÷ 10
24 27 × 10
25 280 × 10
26 570 ÷ 10
27 900 × 10

E Solve these word problems.

28 Cookies are packed in boxes of 10. How many boxes can be filled with 240 cookies?

29 Gus has 23 sets of monster cards. There are 10 cards in each set. How many does he have altogether?

30 Drawing pins come in boxes of 100. How many drawing pins in 46 boxes?

31 A chef orders 100 packs of mints for a restaurant. There are 25 mints in each pack. How many mints altogether?

32 Apples are sold in bags of 10. How many apples in 64 bags?

Challenge

Answer these questions. Write the answers in order, smallest first.

a 27 × 10
b 2600 ÷ 10
c 72 × 100
d 38 × 100
e 83 × 10
f 160 × 10
g 81 × 100
h 51 × 10
i 15 × 100
j 8200 ÷ 10

Can you multiply 2-digit numbers by 100?

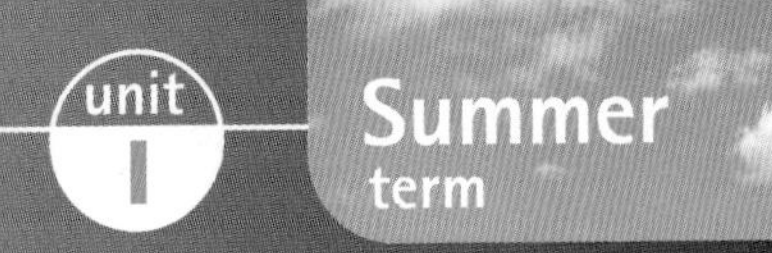

A Write the number at the mid-point of each number line.

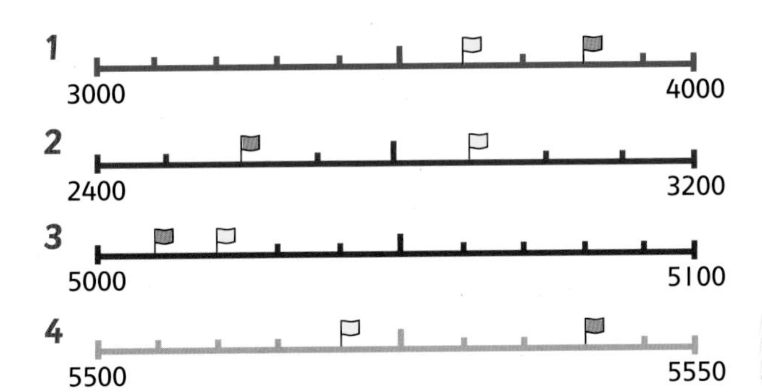

B Write the number shown by the pink flag on each colour number line.

5 3800

5 blue
6 red
7 purple
8 orange

C Write three different numbers that lie between the pink and yellow flags on each number line.

9 3614, 3702, 3729

9 blue
10 red
11 purple
12 orange

D Write the number halfway between each pair of flags.

13 260 and 290
14 320 and 370
15 290 and 330
16 2900 and 3000
17 3270 and 3290
18 5050 and 5100
19 990 and 1010

E Write a number to make each statement true.

20 3600 > ● > 3550
21 2791 < ● < 3021
22 4236 > ● > 1111
23 7032 > ● > 6999
24 5409 < ● < 5904
25 2222 > ● > 1122

F Write the greatest and least amounts that each missing number could be to make these statements true. Use whole pounds.

26 greatest £6469, least £6421

26 £6420 < ● < £6470
27 £1634 > ● > £1592
28 £4123 < ● < £4132
29 £4321 > ● > £1234

Challenge

Use only the digits 5, 7 and 9. Write one digit for each card to make these statements true.

a □□ + □ > □□

b □□ − □ < □□

c □ × □ < □□

d □□□ < □□□ < □□□

A Round each number to the nearest 10.

1 257
2 312
3 879
4 782
5 106
6 111
7 68
8 95
9 105
10 475

B Round each number to the nearest 100.

11 464
12 479
13 303
14 551
15 806
16 279
17 835
18 796
19 405
20 325

C Round the heights of the mountains to the nearest 10 m.

21 Kilimanjaro 5889 m
22 Mont Blanc 4819 m
23 Ben Nevis 1345 m
24 Everest 8856 m
25 Elbruz 5657 m
26 Mount McKinley 6189 m

21 5890 m

D Round the length of the rivers to the nearest 100 km.

27 Danube 2858 km
28 Amazon 6570 km
29 Orinoco 2449 km
30 Nile 6695 km

E Round the numbers in each calculation to the nearest 10 to find an approximate answer. Work out the exact answer.

31 84×12
32 59×13
33 67×15
34 93×14

Challenge

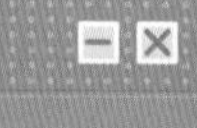

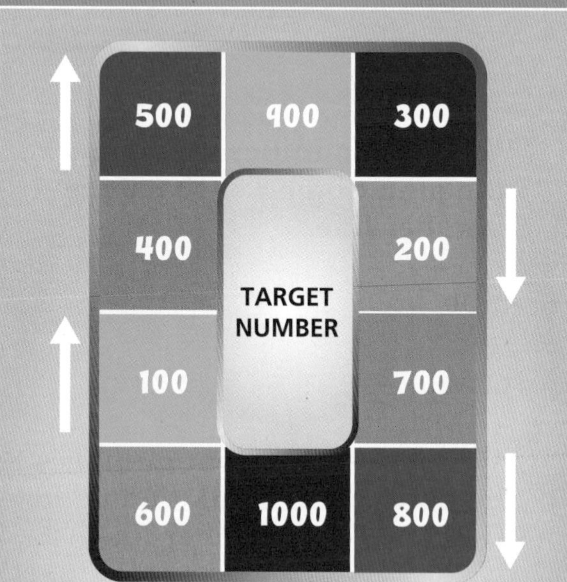

Play with a partner. Place a counter on any target number.
Shuffle number cards 0 to 9 and deal three cards each.
Arrange your cards to make a number as close to the target number as possible. Round your number to the nearest 10. The player closer to the target number scores 1 point. Move the counter forward one section. Play again. The first player to reach 10 points wins.

Can you round any whole number less than 1000 to the nearest 10 or 100?

A Complete these.

1 35 + 9
2 27 + 19
3 54 + 11
4 37 − 19
5 63 + 21
6 92 − 41
7 62 − 19
8 36 + 29
9 53 − 31

B Complete each sequence.

10 51, 60, 69

10 15, 24, 33, 42, ___, ___, ___
11 16, 22, 28, ___, ___, 46, ___
12 27, 38, ___, ___, 71, 82, ___
13 1, 22, 43, ___, ___, 106, 127
14 74, 83, 92, ___, ___, ___, 128
15 65, 76, 87, ___, ___, ___, 131
16 50, 69, 88, ___, ___, 145, ___

C Copy and complete.

17 46 + 19
18 37 + 19
19 53 + 18
20 64 − 19
21 34 − 17
22 181 − 27
23 176 + 27
24 258 + 39
25 374 − 29
26 563 − 28

D Write the total cost for each.

27 pencil and eraser
28 pen and sharpener
29 ruler and eraser
30 pencil and ruler
31 comic and ruler
32 sharpener and pencil

Challenge

Write addition and subtraction problems to complete the clues for the number crossword.

across
a 17 + 9
b
c
d
f
g
h

down
a
b
c
d
e 58 − 21
f
g

a 2	6		b 4	3
8		c 7	6	
	d 9	4		e 3
f 6	2		g 7	7
9		h 4	4	

Can you add or subtract 2-digit numbers ending in 9, 8, 7 or 1 by adding or subtracting the nearest multiple of 10 and adjusting?

A Answer each set of questions. Match each answer to a letter to find who took the cheese.

1 51 T

1 37 + 14
2 45 + 36
3 93 − 26
4 48 + 28
5 26 + 49
6 19 + 72
7 58 − 29
8 71 − 25
9 43 + 48
10 59 + 37

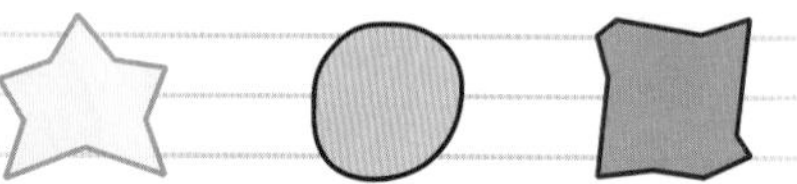

11 49 + 330
12 62 + 120
13 37 + 190
14 19 + 280
15 260 + 43
16 190 + 56
17 29 + 180
18 48 + 360
19 43 + 260
20 38 + 370

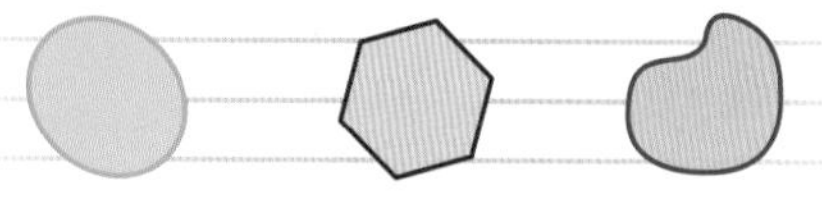

21 33 + 47 + 16
22 26 + 31 + 25
23 15 + 22 + 14
24 18 + 46 + 17
25 25 + 23 + 19
26 54 + 16 + 17
27 19 + 18 + 27
28 23 + 32 + 18
29 38 + 15 + 29
30 26 + 27 + 14
31 28 + 17 + 29

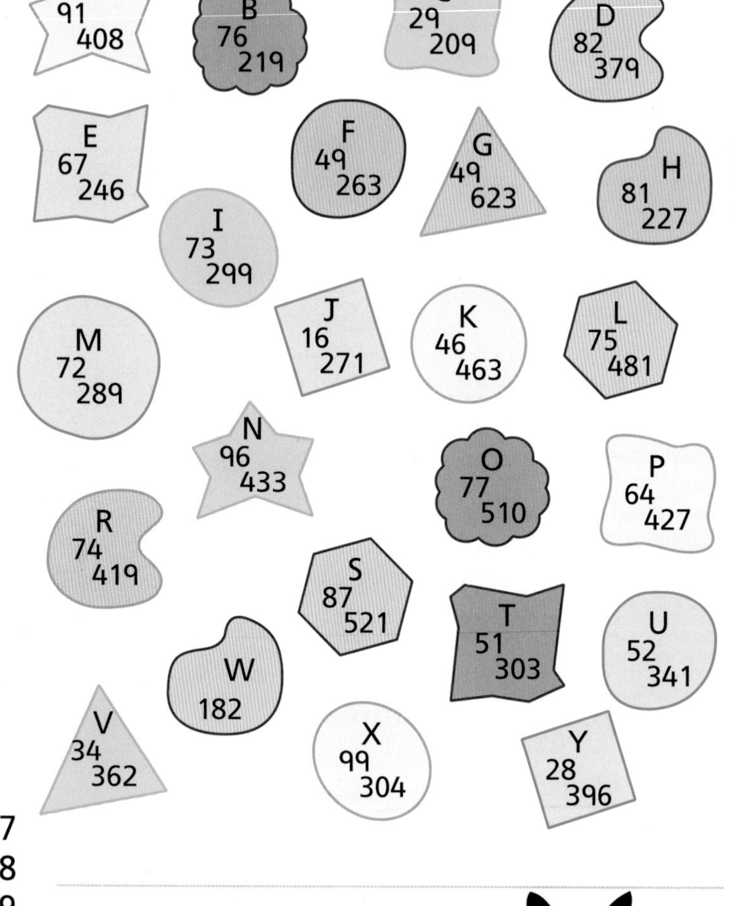

Challenge

One of the other animals saw who took the cheese. Write some addition and subtraction clues for a partner to follow to work out who saw the thief.

Can you add or subtract any pair of 2-digit numbers mentally?

A Choose a method to answer these questions. If you use a written method, show your working.

1. 23 + 494
2. 37 + 395
3. 28 + 473
4. 149 + 56
5. 326 + 123
6. 419 + 387
7. 246 + 664
8. 287 + 549
9. 386 + 368

B Find the total of each set of numbers.

10. 57 62 93 63 124
11. 48 19 72 56 28 147
12. 39 174 58 43 163 72
13. 38 124 69 191 86 186
14. 101 134 72 111 29 54

C Check Darren's homework. Find which five questions he got wrong. Write ✓ or ✗ for each.

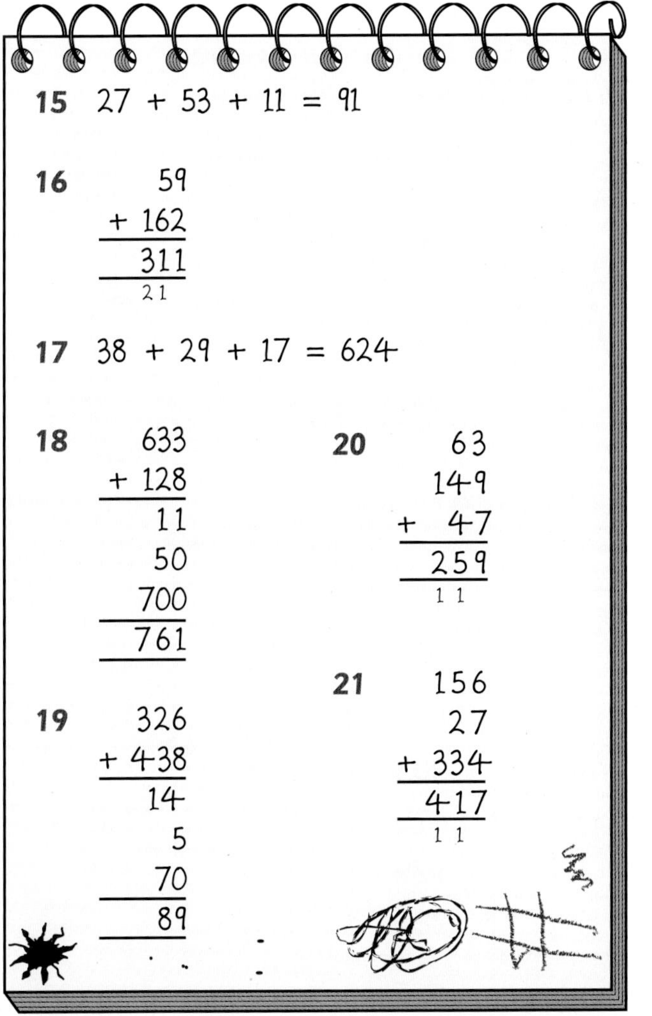

15 27 + 53 + 11 = 91

16
```
   59
+ 162
  311
   21
```

17 38 + 29 + 17 = 624

18
```
  633
+ 128
   11
   50
  700
  761
```

19
```
  326
+ 438
   14
    5
   70
   89
```

20
```
   63
  149
+  47
  259
   11
```

21
```
  156
   27
+ 334
  417
   11
```

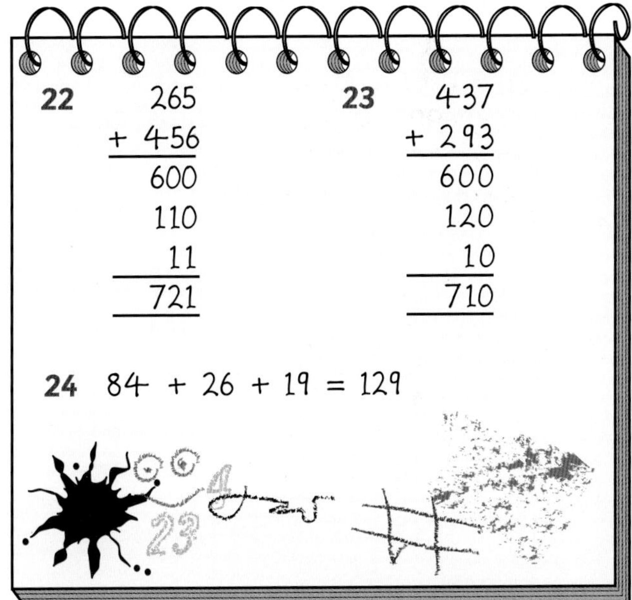

22
```
  265
+ 456
  600
  110
   11
  721
```

23
```
  437
+ 293
  600
  120
   10
  710
```

24 84 + 26 + 19 = 129

Challenge

Darren is not well today. Write an email to him to explain how he made his mistakes.

A Solve these word problems.

1. Dipti has 37 beads. She loses 13. How many are left?
2. When 12 chocolates are taken from a full box, there are 22 left. How many chocolates in a full box?
3. Lollies cost 65p. What is the cost of 3 lollies?
4. Four turns on a fair ride cost £2·20. What is the cost of 1 turn?
5. In a bag of counters, 17 are yellow, 22 are blue, and 15 are green. How many counters altogether?
6. Jack has 44p. Jill has £1·22. If they share their money equally, how much do they have each?

B Write how much money each person has left.

7. Ruby buys a ring
8. Lorne buys a mower
9. Doug buys a fork and spade
10. Russell buys a leaf blower
11. Rose buys a tree
12. Cliff buys a telescope
13. Teli buys a television

Challenge

Take turns with a partner. Roll two dice. Add your two dice numbers together. Find a number in the grid to match the dice total. Repeat to find a second number.
Find the difference between your two numbers. The player with the greater difference scores 1 point. Play again. The first player to 10 points wins.

dice total	number	dice total	number
2	327	8	199
3	421	9	386
4	963	10	678
5	476	11	825
6	129	12	528
7	281		

Can you use a written column method to find the difference between two 3-digit numbers?

A Work out the cost of each pair of fireworks.

1 £7·65

1 shooting star and Roman candle
2 banger and sparkler
3 rocket and banger
4 Catherine wheel and spray
5 shooting star and Catherine wheel
6 spray and shooting star

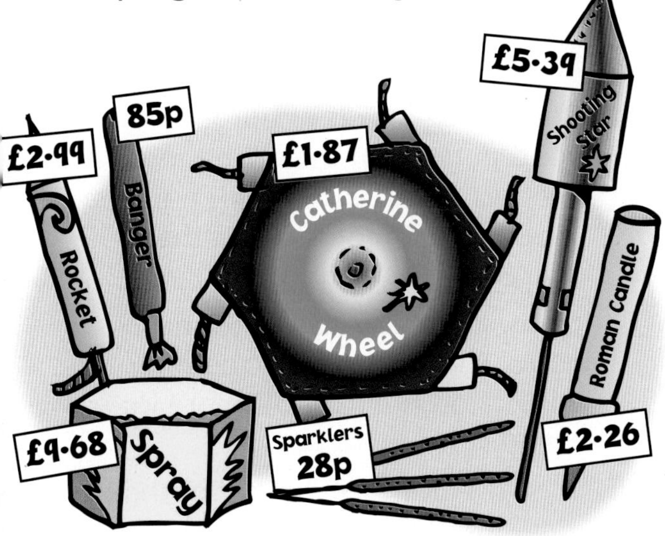

B Find the difference between the prices of each pair.

7 spray and Catherine wheel
8 banger and Catherine wheel
9 rocket and Roman candle
10 shooting star and spray
11 Roman candle and Catherine wheel
12 banger and spray
13 shooting star and rocket

C Answer these questions. Show how you can check your answer using a different method.

14 39 + 46
15 362 + 159
16 241 − 126
17 407 + 248
18 623 − 139
19 176 − 88
20 27 + 58 + 13
21 361 − 163

Challenge

Copy and complete these targets.

Find the missing amounts on each coloured section.

Make your own target with £100 in the centre circle.

a

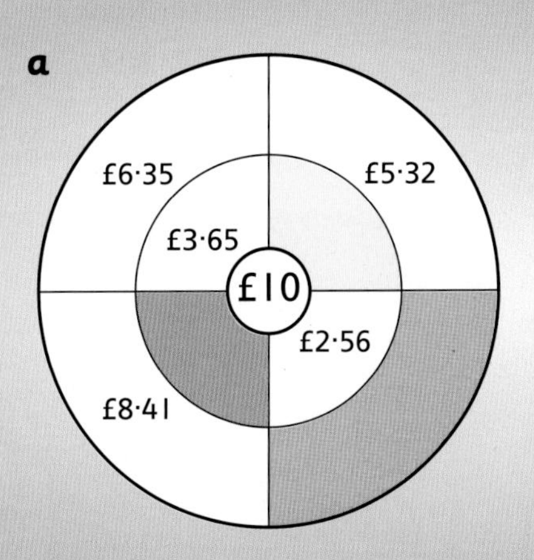

b

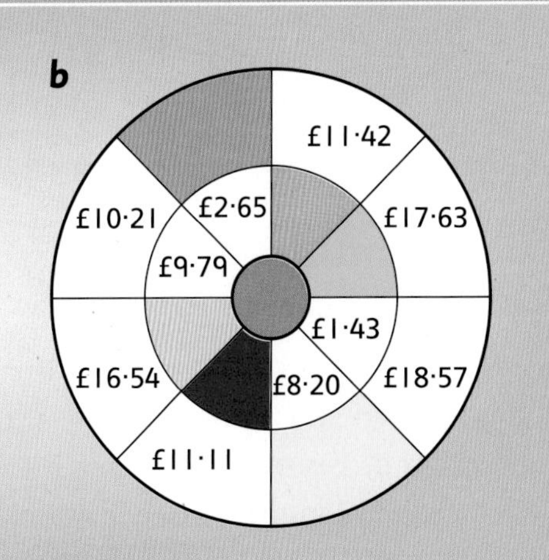

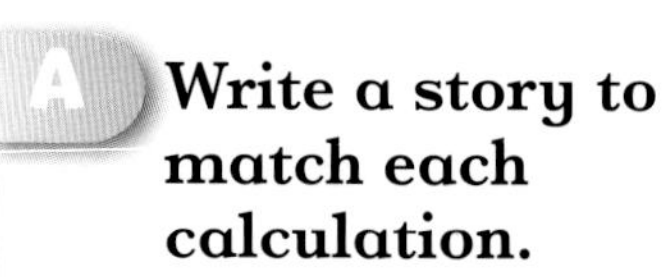

A Write a story to match each calculation.

1 Cars have 4 wheels. How many wheels on 6 cars?

1 $6 \times 4 = 24$
2 $8 \times 5 = 40$
3 $64 \div 8 = 8$
4 $25 + 76 = 101$
5 $123 - 85 = 38$
6 $37 + 120 = 157$
7 $111 - 29 = 82$
8 $4 \times 12 = 48$
9 $55 \div 5 = 11$

B Solve these word problems.

10 In a school there are 217 children. 128 are girls. How many are boys?

11 3 coaches each carry 43 passengers. How many passengers altogether?

12 Liz plants 92 tulip bulbs. 75 tulips grow. How many bulbs do not grow?

13 862 people visited a museum in August. 146 fewer people visited in September. How many people visited in September?

14 Mr Fulofun bought 4 garden gnomes for £3·55 each. How much change did he get from £20?

15 There are 4 boxes of burgers for a party. 3 of the boxes hold 12 burgers and 1 box holds 9 burgers. How many burgers altogether?

16 Kelli has three £1 coins, two 50p coins, three 20p coins and four 5p coins. She buys an ice-cream for £1·65. Can she afford 2 more ice-creams?

17 Kirsty is given £50 for her birthday. How much more does she need to buy trainers costing £29·50, a T-shirt for £7·85 and a jumper for £18·20?

£29·50

Challenge

The numbers in circles are the answers to the sums or differences of two card numbers. Work out the questions to match each answer.

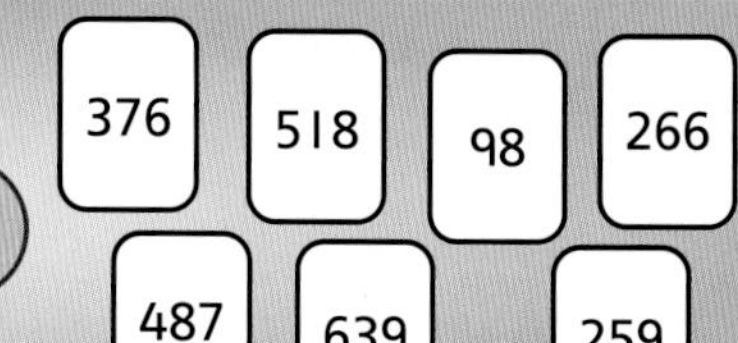

Can you use all four operations to solve word problems?

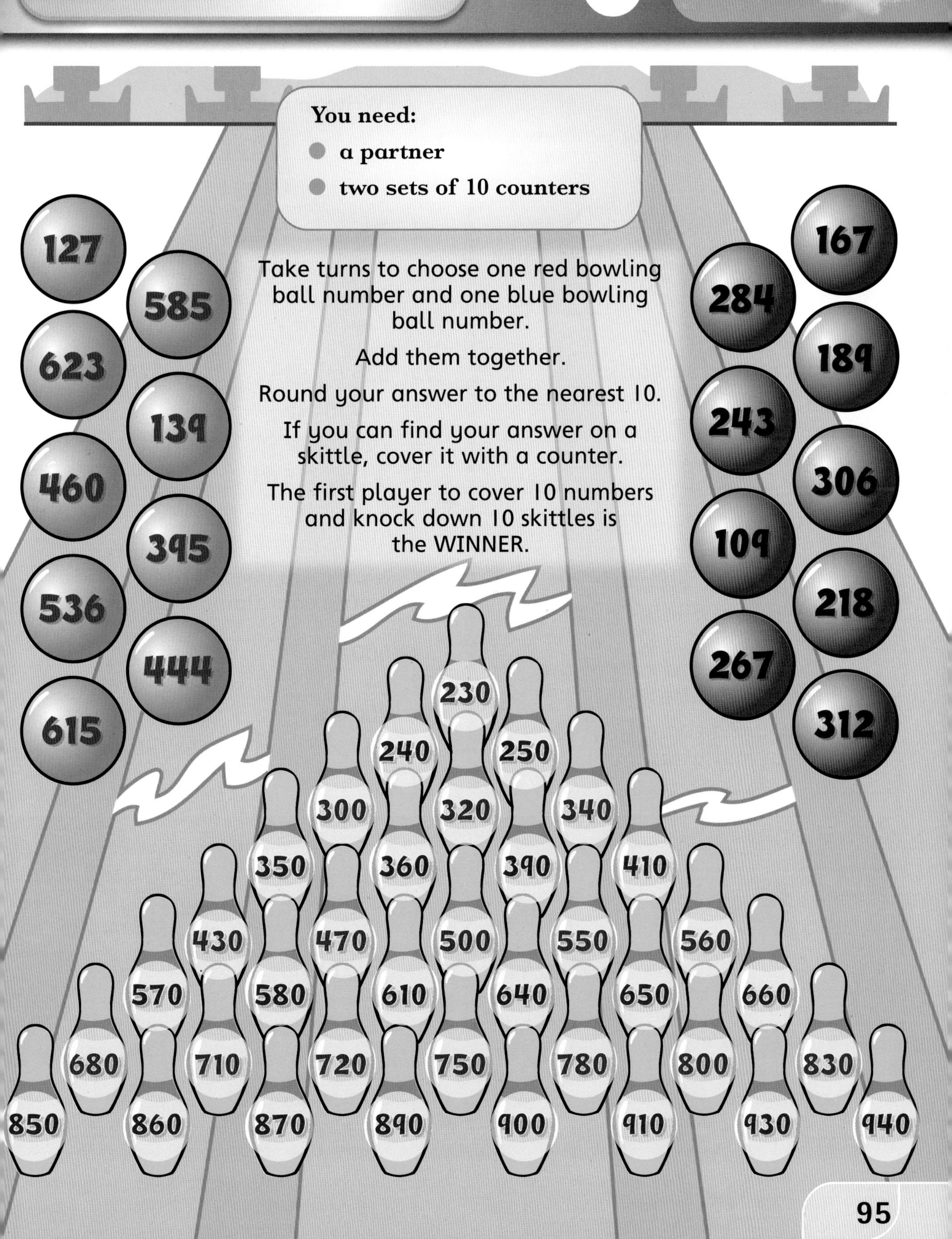
You need:
a partner
two sets of 10 counters
Take turns to choose one red bowling ball number and one blue bowling ball number.
Add them together.
Round your answer to the nearest 10.
If you can find your answer on a skittle, cover it with a counter.
The first player to cover 10 numbers and knock down 10 skittles is the WINNER.
127
585
623
139
460
395
536
444
615
167
284
189
243
306
109
218
267
312
230
240
250
300
320
340
350
360
390
410
430
470
500
550
560
570
580
610
640
650
660
680
710
720
750
780
800
830
850
860
870
890
900
910
930
940

A Solve these capacity problems.

1. How many full glasses of water would fill the bottle?
2. How many full glasses of water would fill the mug?
3. How many full mugs of water would fill the bucket?
4. If you fill the glass and the mug from a full bottle of water, how much water is left in the bottle?
5. What is the capacity of the mug in millilitres?
6. What is the capacity of the vase in millilitres?

B Write in millilitres.

7 $\frac{1}{10}$ l
8 $\frac{1}{2}$ l
9 $\frac{1}{4}$ l
10 2 l
11 $\frac{3}{4}$ l
12 $3\frac{1}{2}$ l

7 100 ml

C Write the missing sign for each. Use < or >.

13 526 ml ● $\frac{1}{2}$ l
14 200 ml ⬣ $\frac{1}{2}$ l
15 700 ml ✸ $\frac{3}{4}$ l
16 $\frac{3}{10}$ l ◖ 325 ml
17 $1\frac{1}{2}$ l ▲ 1550 ml
18 1500 ml ✹ $1\frac{1}{4}$ l

13 526 ml > $\frac{1}{2}$ l

D Write how many millilitres of water are needed to fill a litre bottle containing each amount.

19 400 ml
20 $\frac{1}{4}$ l
21 200 ml
22 $\frac{3}{10}$ l
23 $\frac{9}{10}$ l
24 $\frac{3}{4}$ l
25 720 ml
26 $\frac{1}{2}$ l
27 190 ml

Challenge

Play with a partner. Each place a counter on START. Take turns to roll a dice and move around the pond. Keep a record of how much water you collect. The first player to collect 10 l wins.

Do you know the equivalent of $\frac{1}{2}$, $\frac{1}{4}$, $\frac{3}{4}$ and tenths of 1 litre in millilitres?

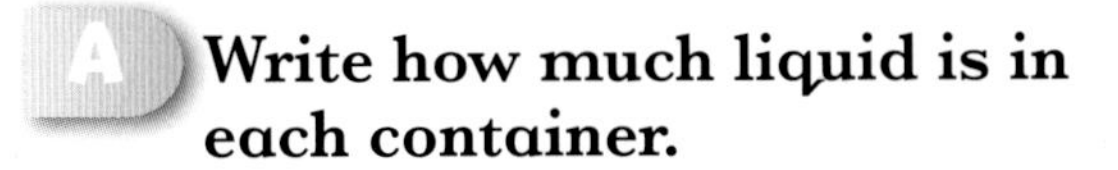

A Write how much liquid is in each container.

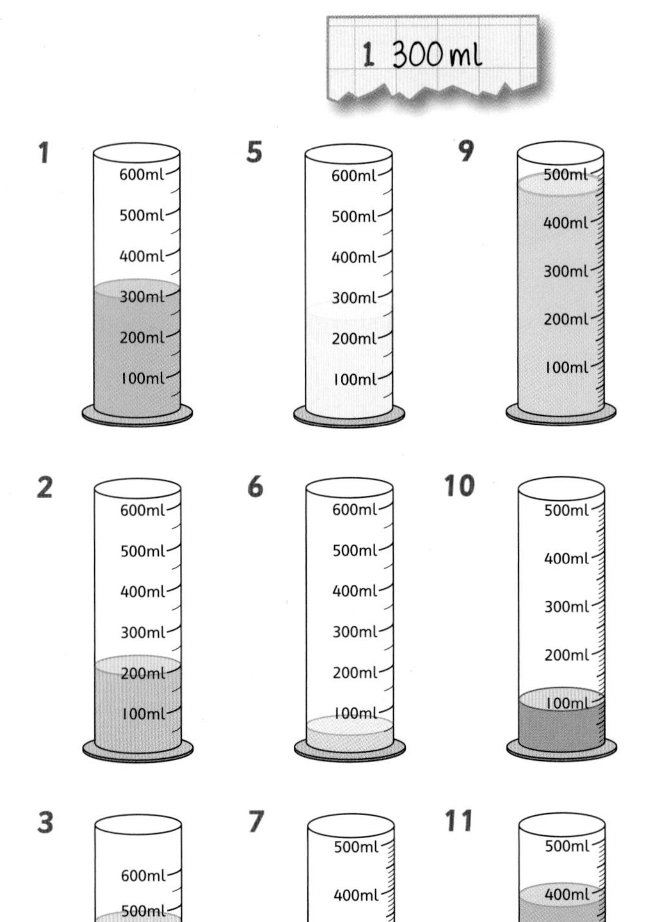

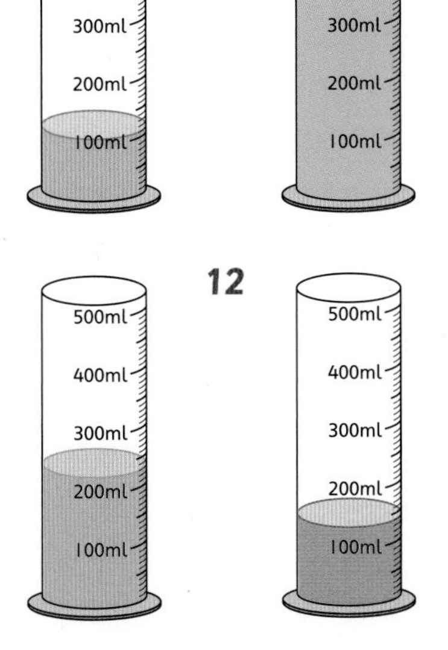

B Write how much water is in each container.

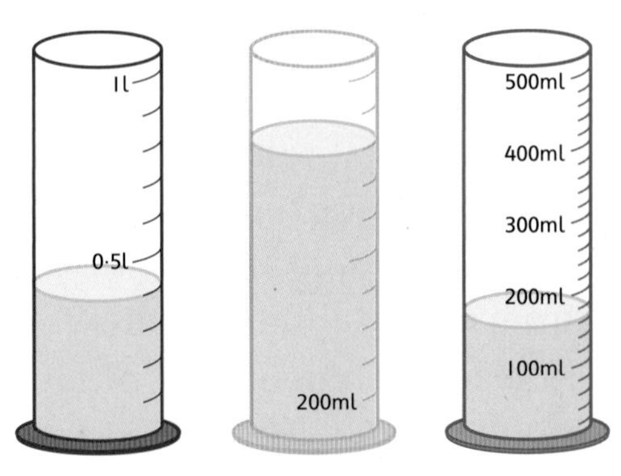

13 the red container
14 the yellow container
15 the blue container
16 the red container when half full
17 the yellow container when half full
18 the blue container when $\frac{3}{10}$ full
19 Write how much more water is needed to fill each container.

13 400 ml

Challenge

Explain how to use these containers to measure out each amount of liquid.

a 180 ml **d** 1·2 l
b 190 ml **e** 3300 ml
c 200 ml **f** 2 l

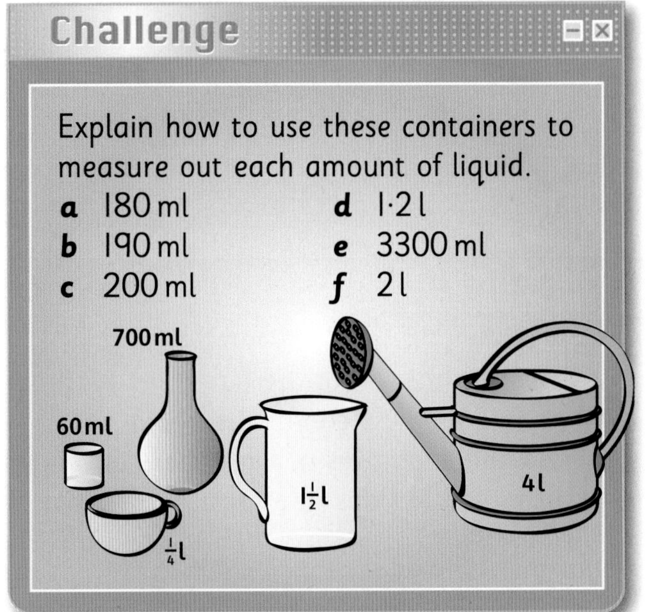

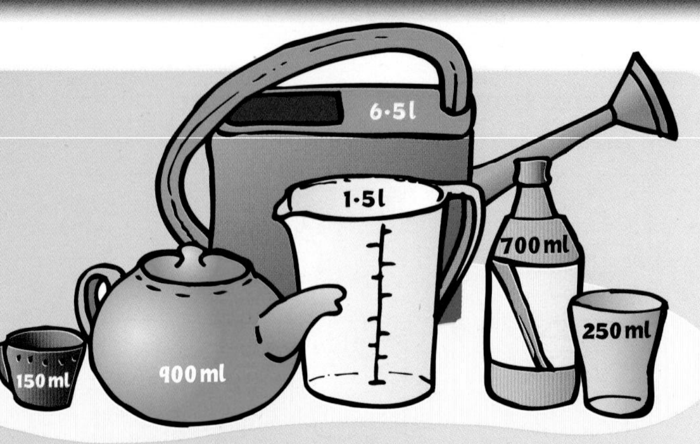

A Solve these capacity problems.

1. How many cups can you fill from a full teapot?
2. How many glasses can you fill from a full jug?
3. How many cups can you fill from a bottle full of water?
4. How many teapots full of tea are needed to fill 24 cups?
5. If there is $\frac{1}{2}$ l of water in the bottle, how much is needed to fill it up?
6. How many times can you fill a bottle with a full watering can?
7. The teapot, bottle and another container can hold $3\frac{1}{2}$ l of water altogether. What is the capacity of the third container?
8. Which containers above hold more than $\frac{1}{2}$ l but less than 1 l?
9. How many glasses full of water are needed to fill 10 cups?
10. If 300 ml is poured from a full jug, how much water is left in the jug?
11. A cup and glass are filled from a full jug. How much liquid is left in the jug?
12. Can a watering can full of water fill all the other containers twice?
13. If 15 ml of each cup of tea is milk and the rest is hot water, how much hot water is needed for 12 cups of tea?
14. A tin of soup contains 400 ml. When heated, 100 ml of milk is added to each can. How much milk is needed to make 3 l of soup mixture?
15. A tin of soup costs 68p and milk costs 42p a litre. What is the cost of making 30 l of the soup mixture in question 14?

Challenge

Copy and complete the bar chart to show the capacities of the five smallest containers above.
Find the capacities of five other containers that hold less than 1 litre. Display this information on a bar chart.

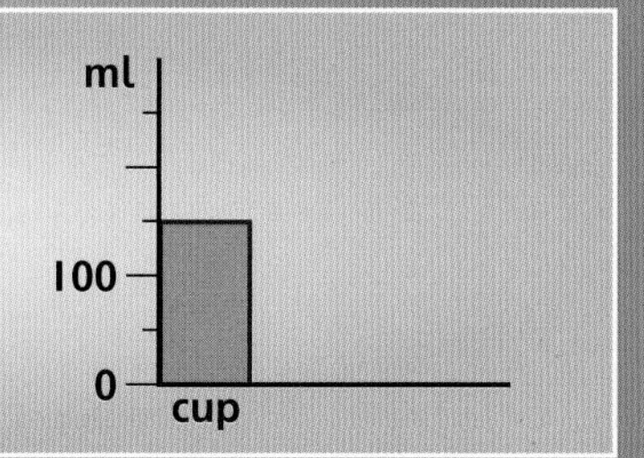

Can you use all four operations to solve word problems involving capacity?

A Write the colour of the line of symmetry for each shape.

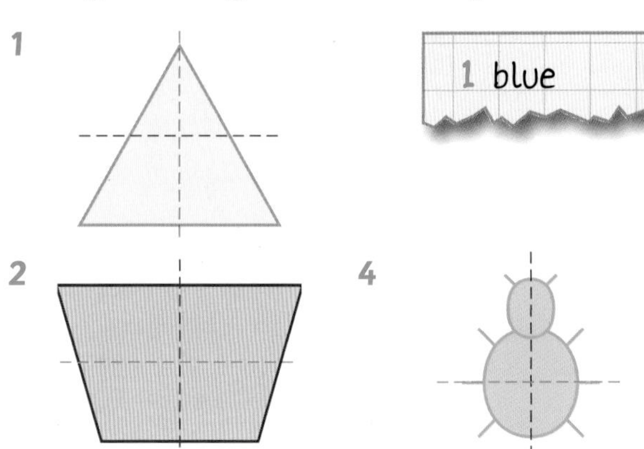

C Find whether each shape has 1, 2 or no lines of symmetry.

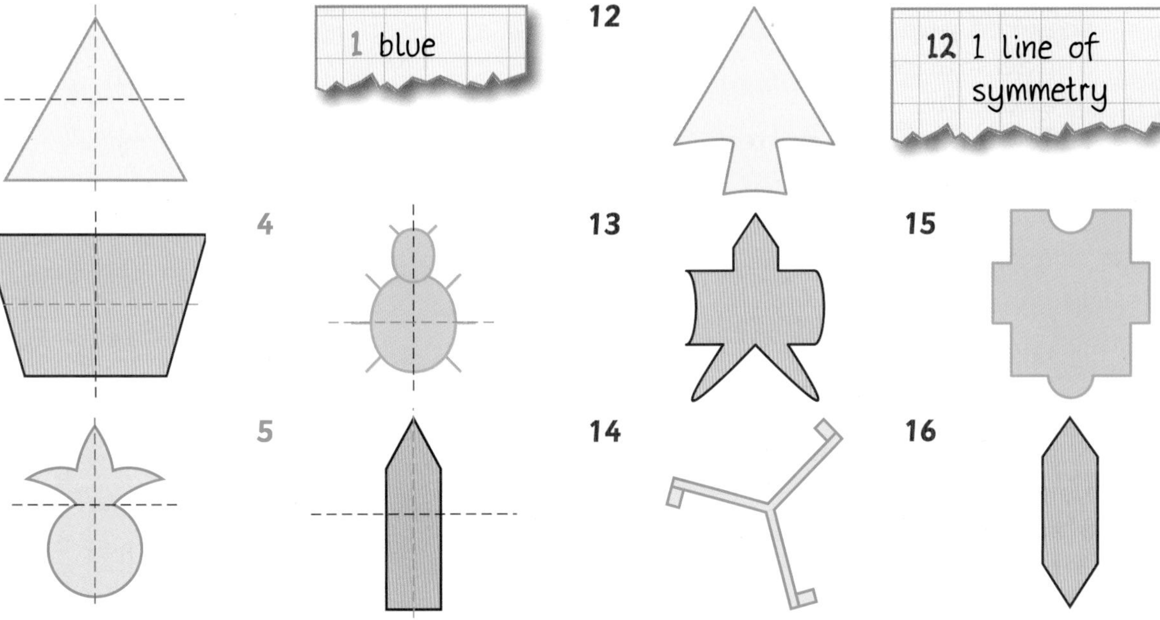

B Write how many lines of symmetry you can find for each shape.

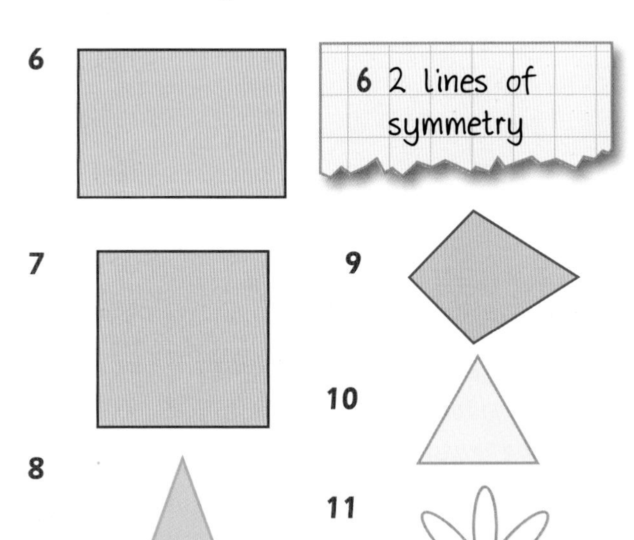

D Answer these questions.

A B C D E F G H I J K L M

N O P Q R S T U V W X Y Z

17 Which letters have line symmetry?

18 Draw one line of symmetry for each symmetrical letter.

A **Copy each pattern and mirror line onto squared paper. Shade squares to show the reflection of each pattern.**

1
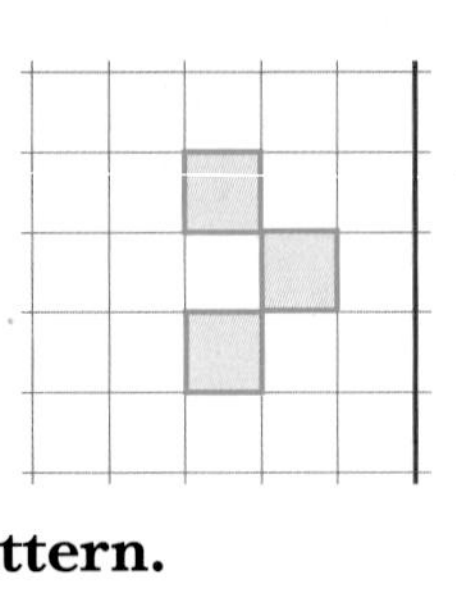

2
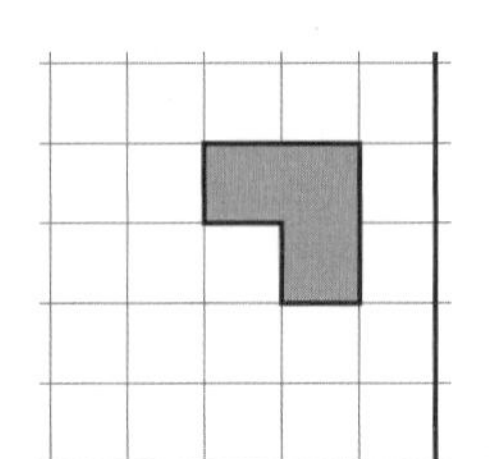

3
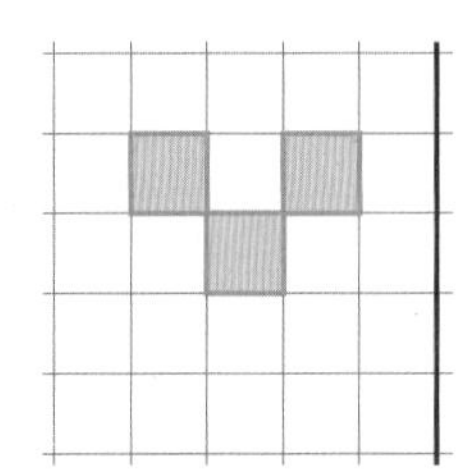

4
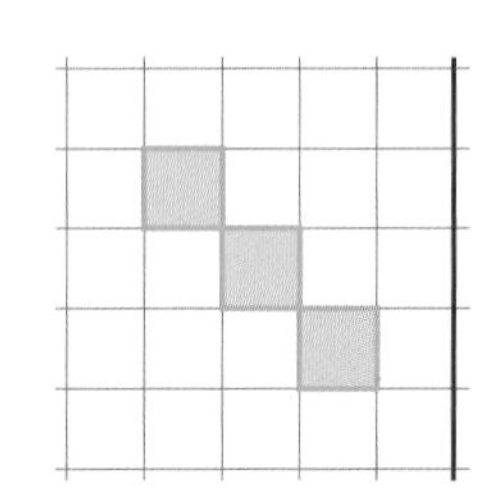

5
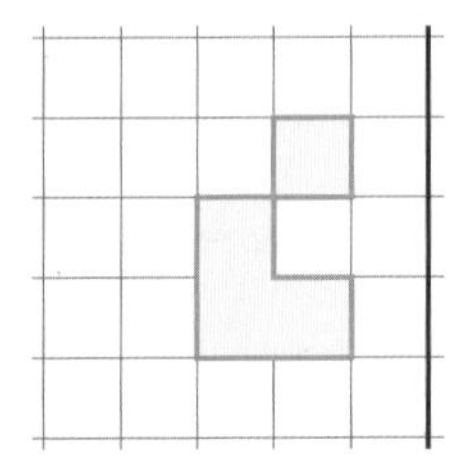

B **Copy each shape and mirror line onto squared paper. Draw each reflected shape.**

6
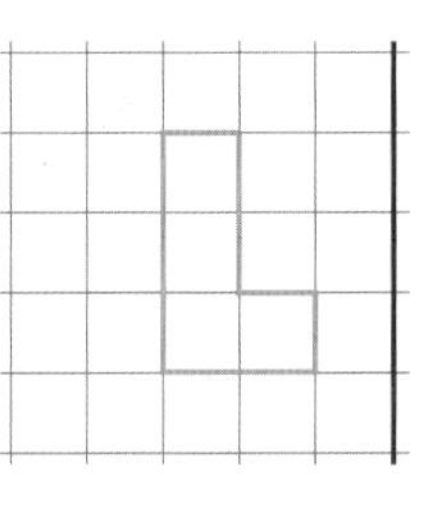

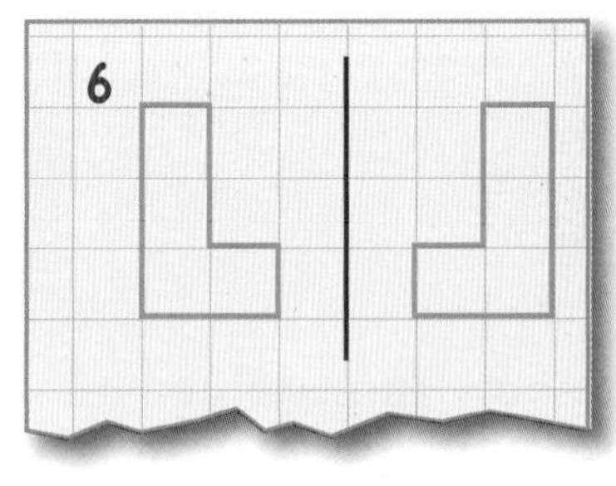

7
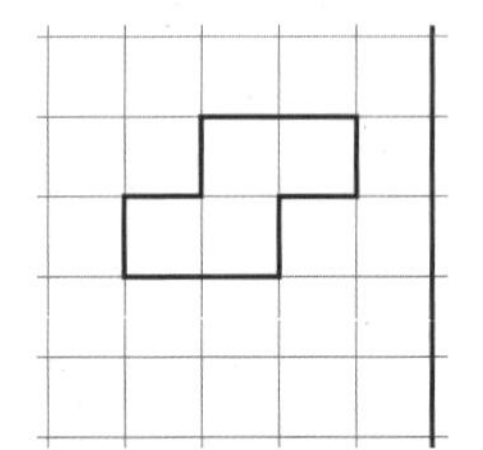

8
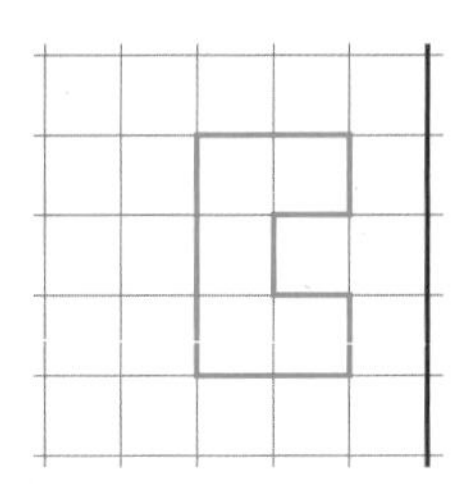

9
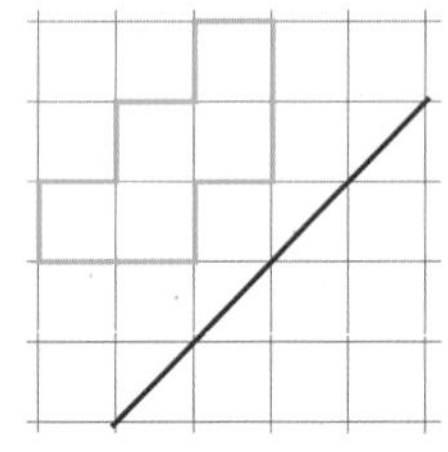

10
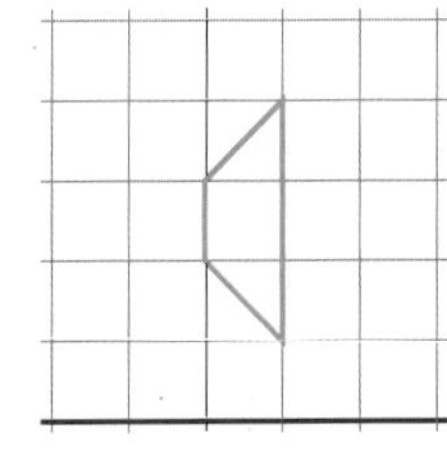

C **Copy each shape onto squared paper. Make patterns of shapes using repeated horizontal or vertical translations.**

11
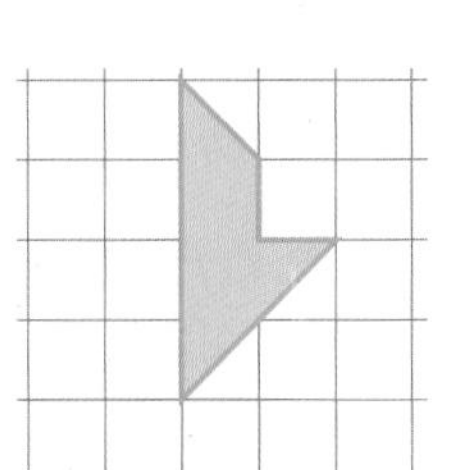

12
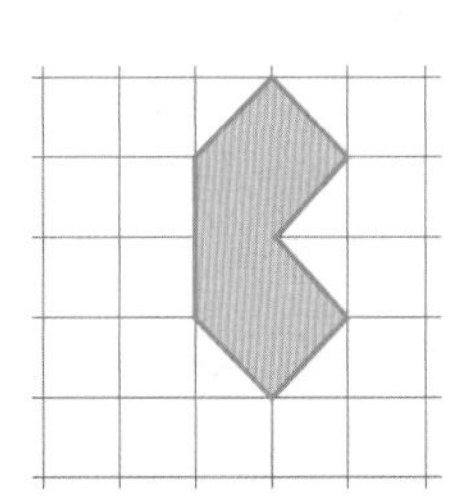

Challenge

Design two bookmarks or CD covers. The first should have a symmetrical pattern. The second should have a pattern formed by translating a shape.

Can you understand and use the terms horizontal, vertical, translation and reflection?

A

Write the co-ordinates of the vertices of triangle A in each position.

1 (4,1) (7,1) (4,6)

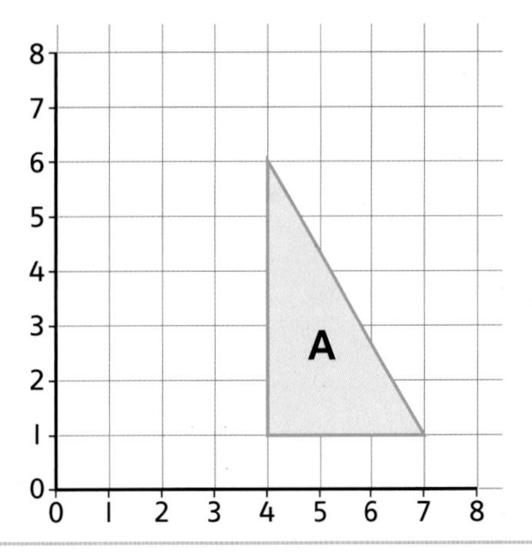

1. as shown on the grid
2. translated 4 squares to the left
3. translated 2 squares vertically upwards
4. translated 1 square right

B

Write the co-ordinates of the corners of rectangle B in each position.

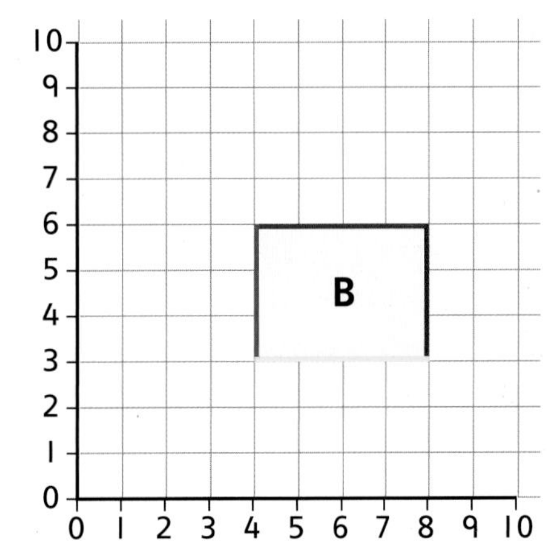

5. as shown on the grid
6. reflected through the red side
7. reflected through the blue side
8. reflected through the purple side
9. reflected through the yellow side

C

Use a co-ordinate grid drawn on squared paper.

10. Draw a triangle with vertices at (2,3), (7,6) and (5,0).
11. Translate the triangle 5 squares to the right. What are the co-ordinates of the vertices?
12. Draw a hexagon with vertices at (5,2), (7,2), (9,5), (7,6), (7,5) and (3,4).
13. Translate the hexagon 2 squares vertically upwards. What are the co-ordinates of the vertices?

Challenge

Use cm-squared paper. Draw an interesting shape. Translate the shape to design a wallpaper border for a room in your home.

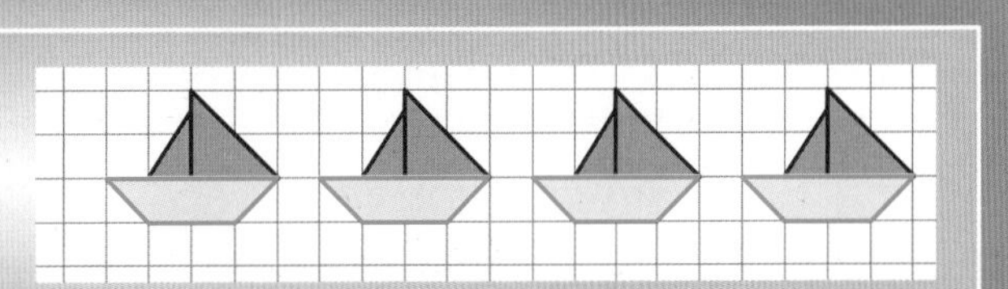

Counter moves

You need: a partner, a dice, coloured counters (three red, three blue, three yellow, three green)

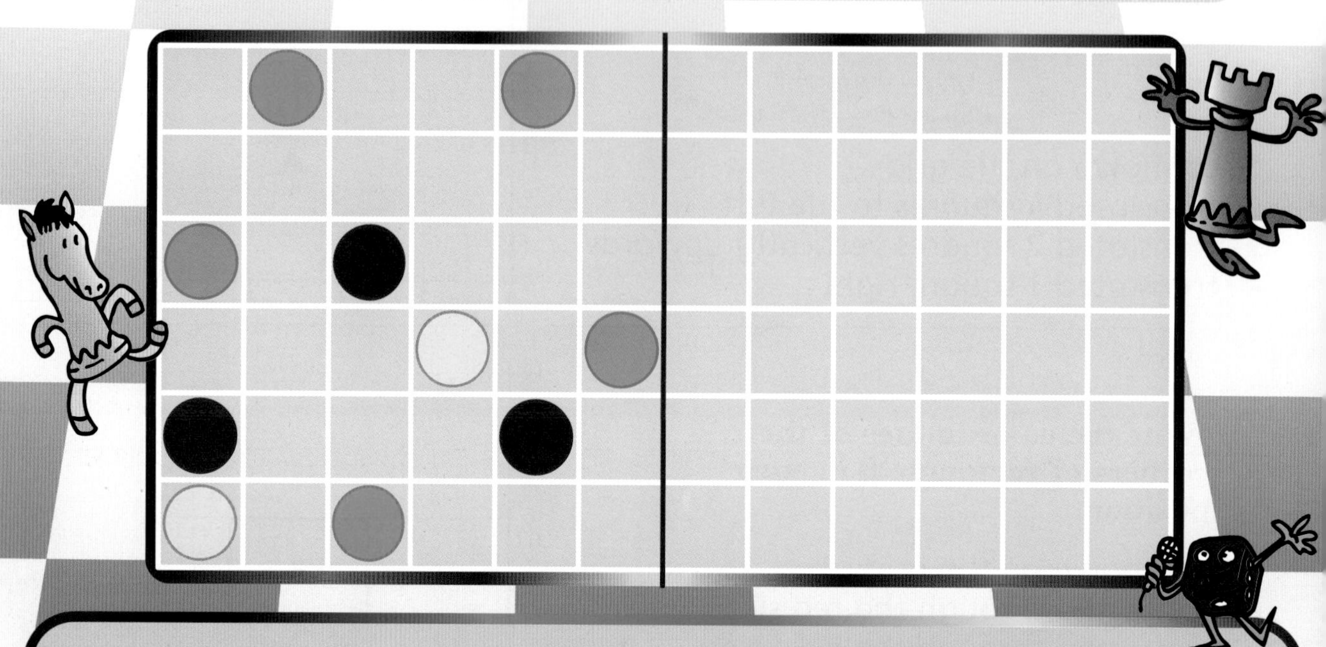

Game A Roll the dice to select a counter.

dice number	counter colour
1 or 2	●
3 or 4	●
5	●
6	○

Place your counter to the right of the mirror line to show the reflection of a counter of the same colour.

Count how many dice throws you need to reflect all the counters.

Can your partner beat your record by taking fewer throws?

Game B Take your 12 counters. You can swap each counter for the amount of water shown in the table.

How much water will you have if you swap all your counters?

How many of each colour counter will you need to swap to have exactly 1 litre of water?

Find 10 other ways you can swap counters to have 1 litre of water.

counter	amount of water
●	50 ml
●	$\frac{1}{4}$ l
○	20 ml
●	100 ml

A Compare each angle to a right angle. Write 'smaller' or 'larger'.

1

5

2

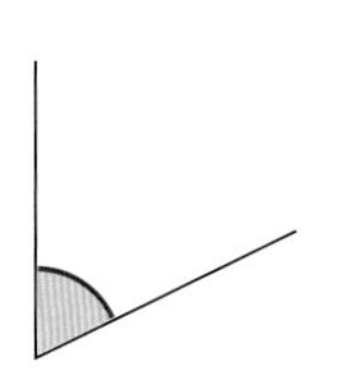

6

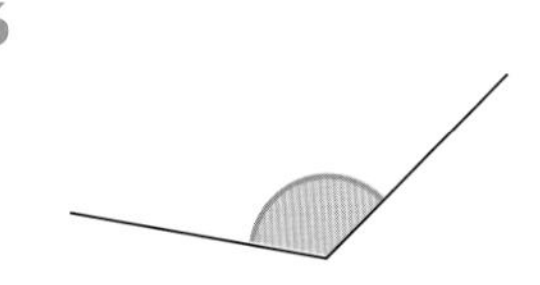

3

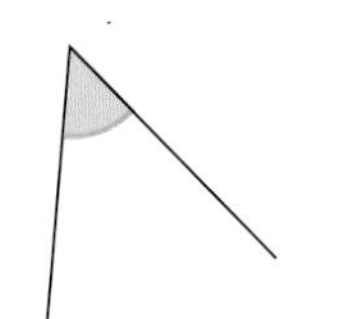

7

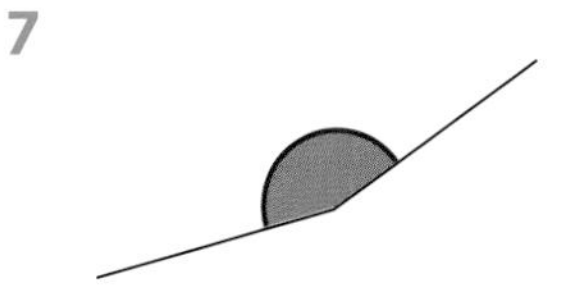

4

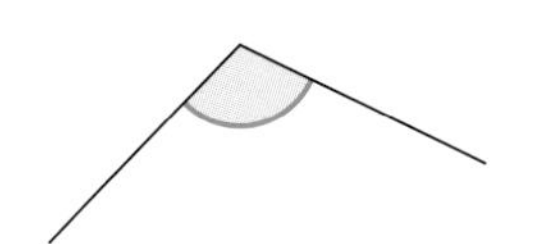

8

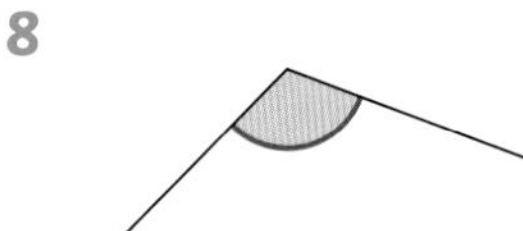

B Write the angle the red hand turns through when it moves clockwise.

9 60°

9 from 3 to 5
10 from 7 to 8
11 from 10 to 12
12 from 1 to 4
13 from 6 to 10
14 from 12 to 3
15 from 5 to 11
16 from 6 to 6
17 from 11 to 3
18 from 1 to 7

C Write the letter the hand points to after each turn.

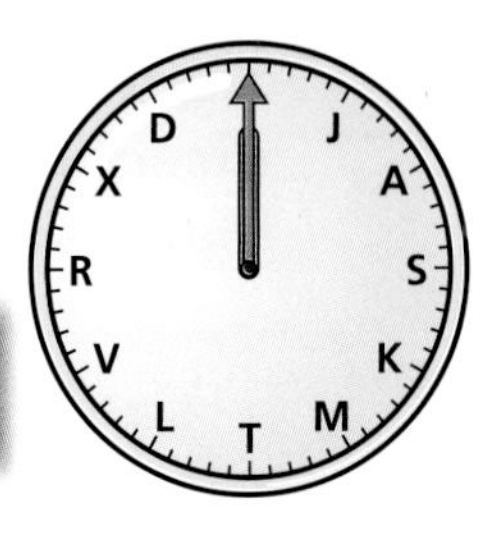

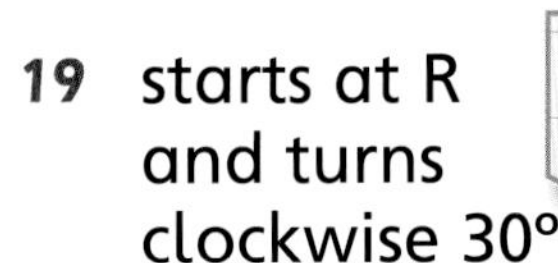

19 X

19 starts at R and turns clockwise 30°
20 starts at J and turns clockwise 60°
21 starts at S and turns clockwise 90°
22 starts at L and turns anti-clockwise 30°
23 starts at X and turns anti-clockwise 180°
24 starts at M and turns clockwise 120°
25 starts at D and turns clockwise 360°
26 starts at T and turns anti-clockwise 150°

Challenge

Use these two clocks to create a secret message. Work out this message first. The hands start pointing vertically upwards each time.

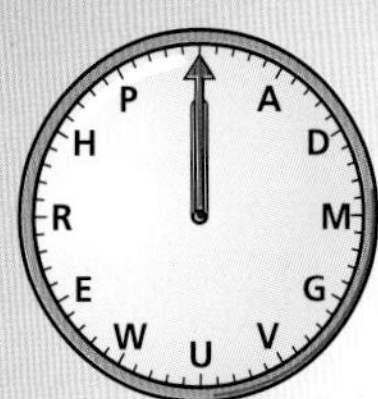

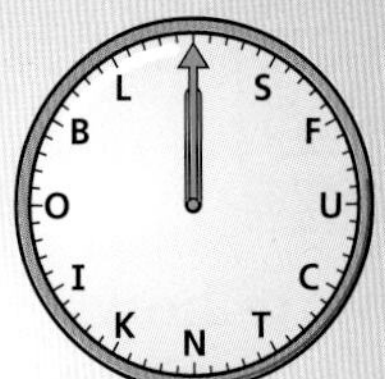

Red hand 120° clockwise	**G**
Blue hand 90° anti-clockwise	**O**
Blue hand 270° clockwise	
Red hand 60° clockwise	
Blue hand 30° anti-clockwise	
Red hand 180° clockwise	
Blue hand 120° clockwise	
Blue hand 150° anti-clockwise	

Can you make turns of 30°, 60° and 90° on a clock face?

A Write the direction each boat is sailing.

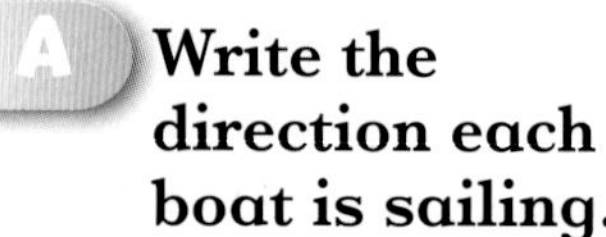

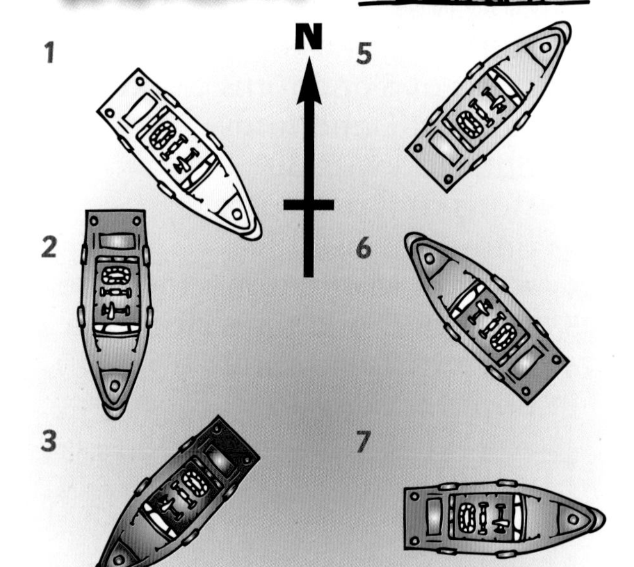

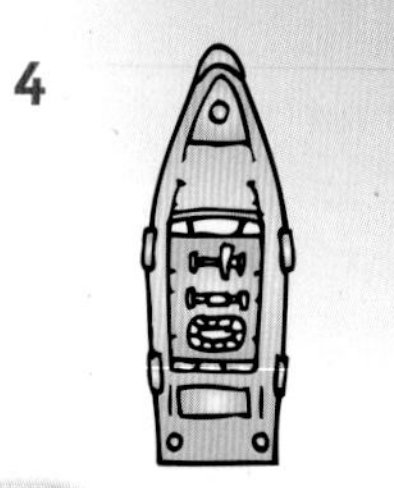

B Write how many right angles each boat turns through. All boats turn clockwise.

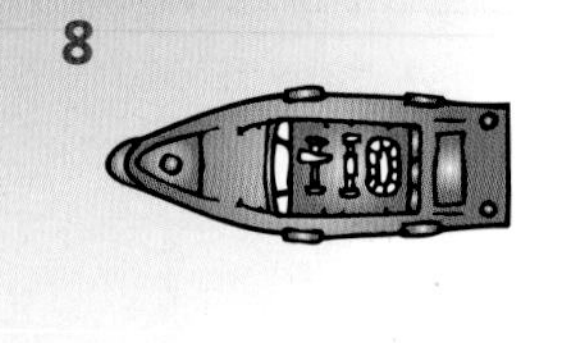

9 pink boat turns to sail SE
10 blue boat turns to sail S
11 orange boat turns to sail N
12 yellow boat turns to sail S
13 brown boat turns to sail E
14 red boat turns to sail E
15 purple boat turns to sail E

C Write the direction each boat is sailing after these turns.

16 blue boat turns 45° clockwise
17 purple boat turns 45° clockwise
18 yellow boat turns 90° anti-clockwise
19 pink boat turns 360° clockwise
20 brown boat turns 135° anti-clockwise
21 red boat turns 180° anti-clockwise
22 orange boat turns 135° clockwise

Challenge

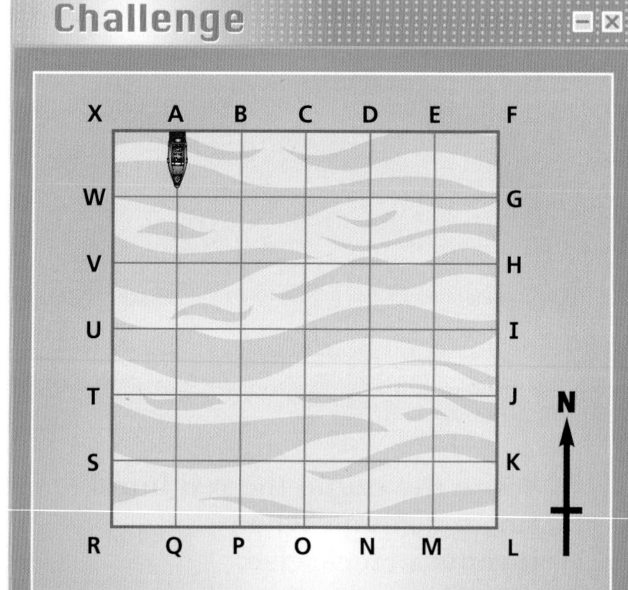

Follow this route. What letter do you arrive at? Clue: You finish after tea!

Start at A. Move 1 section S, 2 sections SE, 1 section N, 2 sections W, and 1 section SW.

Make up your own routes for each journey.

a D to T
b J to C
c M to H
d R to G

Can you make and describe turns using compass directions?

A These angles measure 30°, 45°, 60°, 90°, 120°, or 150°. Estimate the size of each.

1
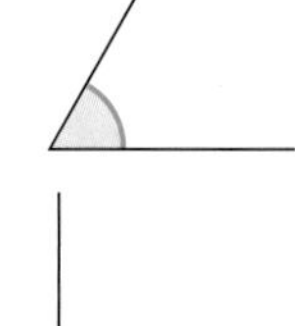

2

3
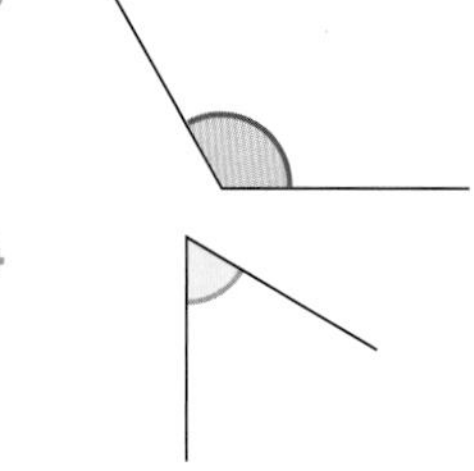

4

5
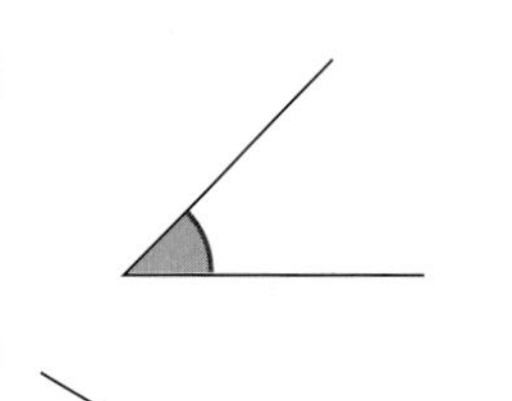

6

7
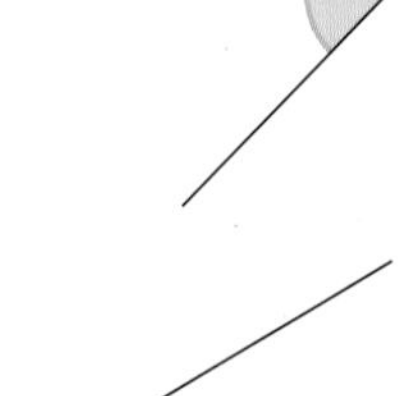

8

B Use the grid to answer these.

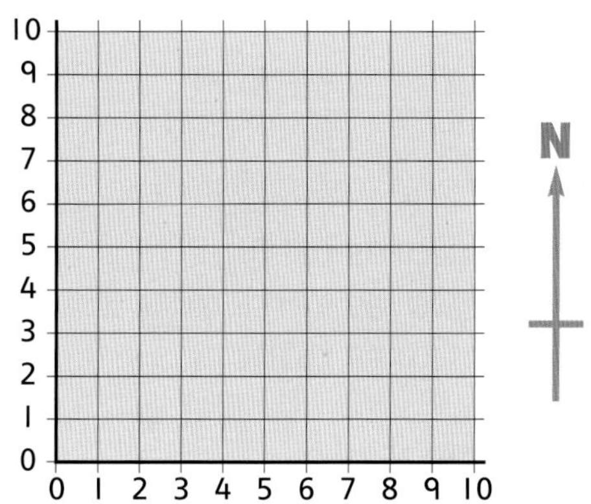

9 (6,3) (6,6)

9 Give 2 points that lie north of (6,1).
10 Give 2 points that lie east of (2,4).
11 Give 2 points that lie south of (0,7).
12 Give 2 points that lie south-west of (9,10).
13 Give 2 points that lie north-west of (10,2).
14 What point is 2 sections south of (4,7)?
15 What point is 3 sections north-east of (1,0)?
16 What point is 4 sections south-west of (5,5)?

C Use a 10 by 10 co-ordinate grid. Write the name of each shape you draw.

17 Start at (2,1). Move 3 sections NE, 3 sections SE, then 6 sections W.
18 Start at (0,10). Move 5 sections E, 5 sections SE, 5 sections W, then 5 sections NW.

D Measure each angle as accurately as possible.

19 25°

19

20
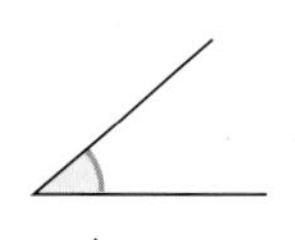

21

22
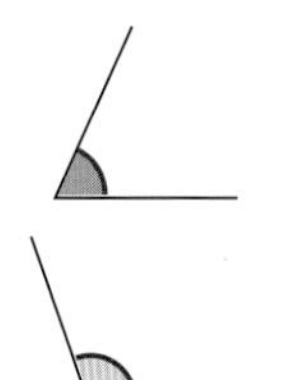

23

Challenge

Create a picture on a grid. Write instructions like those in section C for a partner to follow and draw the picture.

Can you find the position of a point on a grid of squares where the lines are numbered?

Review 5

A Round each length to the nearest 10 m.

1 76 m
2 149 m
3 306 m
4 724 m
5 365 m
6 155 m

B Round each distance to the nearest 100 km.

7 4917 km
8 6246 km
9 3207 km
10 1099 km
11 3650 km
12 1050 km

C Choose a method to complete these.

13 22 + 56 + 14
14 36 + 15 + 29
15 63 + 297
16 326 + 148
17 493 + 489

D Solve these problems.

18 Billy has £48 and spends £19. How much has he left?

19 Ice-creams cost £1·35. What is the cost of 3 ice-creams?

20 Four cinema tickets cost £3·40. What is the price of 1 ticket?

21 On a farm with 118 animals, 36 are cows, 8 are horses, 15 are goats and the rest are sheep. How many sheep are there?

E Write in millilitres.

22 $\frac{1}{2}$ l
23 3 l
24 $1\frac{1}{2}$ l
25 $2\frac{1}{2}$ l
26 $\frac{3}{4}$ l
27 $\frac{1}{4}$ l

F Write the co-ordinates of the vertices of the triangle after each translation.

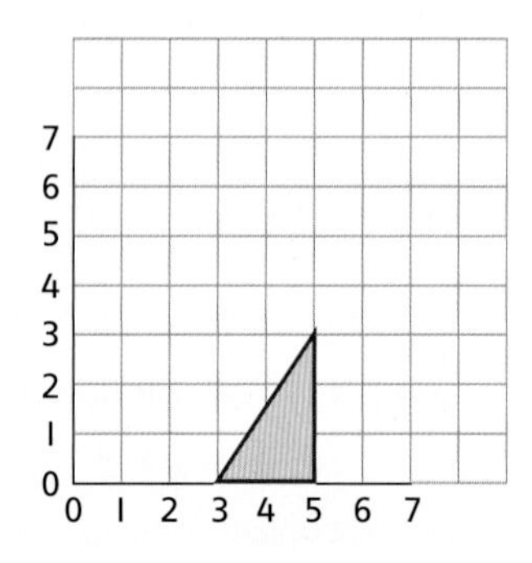

28 3 squares left
29 2 squares vertically upwards
30 2 squares right

G Copy and complete the table.

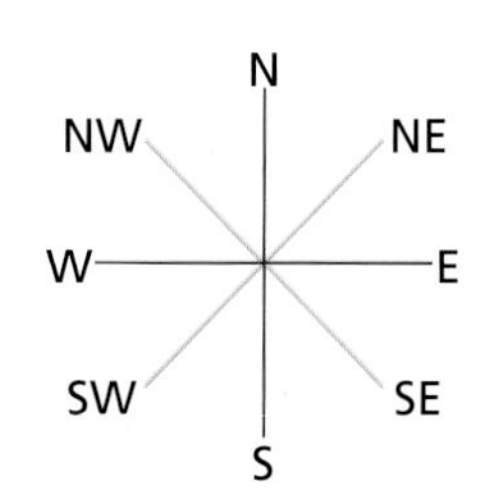

	I am facing	I turn	I am now facing
31	N	90° anti-clockwise	
32	E	45° clockwise	
33	SW	360° anti-clockwise	
34		180° clockwise	S
35	NE	135° anti-clockwise	

A Answer these questions about multiples.

1 25, 40

1. Which numbers on red cars are multiples of 5?
2. Which numbers on yellow cars are multiples of 2?
3. How many cars show a multiple of 10?
4. How many cars show a multiple of 50?
5. If a car with a multiple of 100 wins the race, what colour will the car be?
6. How many cars show a multiple of 2?

B Copy and complete each Venn diagram with the numbers shown.

7 all the car numbers in the picture

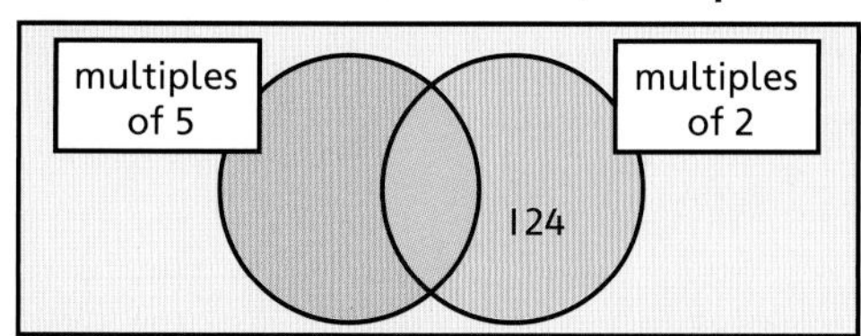

8 numbers 1 to 40

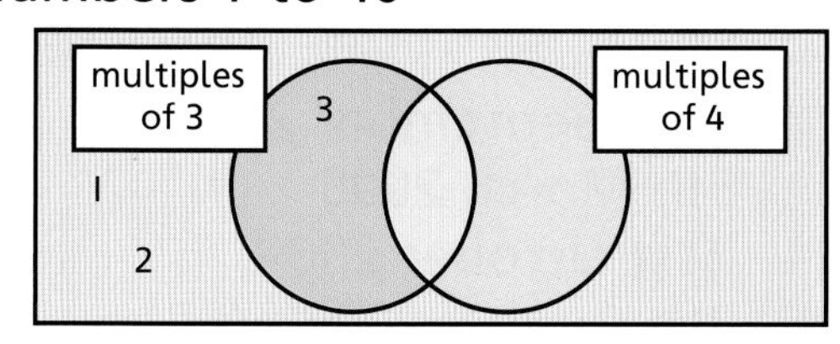

C Write two numbers that are multiples of these numbers.

9 6, 18

9	2 and 3	13	2, 3 and 5
10	3 and 4	14	2, 4 and 10
11	5 and 10	15	3, 4 and 5
12	4 and 5	16	5, 50 and 100

D Use a 100 square. Find which three consecutive numbers have each total.

17 22, 23, 24

17 69
18 42
19 84
20 129
21 243

E Use a 100 square. Find which three numbers vertically next to each other have each total.

22 16, 26, 36

22 78
23 144
24 90
25 126
26 225

Challenge

Use a 100 square. Find as many different totals up to 100 as possible by adding:

a three consecutive numbers
b three numbers written in a vertical line.

A Write two number sentences that contain these numbers.

1 15 ÷ 3 = 5, 3 × 5 = 15

1 3, 5 and 15
2 4, 5 and 20
3 11, 5 and 6
4 8, 16 and 2
5 9, 3 and 27
6 10, 20 and 2
7 5, 6 and 30

B This is the space city Navros. The year is 3003. A spaceship has brought a very old, damaged piece of paper from a distant planet called Earth.

8 Copy and complete the rhyme for the citizens of Navros.

ays hath Septe
ril, june and Nov
he rest have 3
cepting February alone,
hich ha 8 days clear
And 2 each leap yea

C Write how many days in each time period.

9 a week
10 a year
11 June
12 August
13 4 weeks
14 October
15 a leap year

D Look at the Navros calendar for 3003. Write what day each date in 3003 will be.

16 Monday

January 3003

Mo	Tu	We	Th	Fr	Sa	Su
		1	2	3	4	5
6	7	8	9	10	11	12
13	14	15	16	17	18	19
20	21	22	23	24	25	26
27	28	29	30	31		

February 3003

Mo	Tu	We	Th	Fr	Sa	Su
					1	2
3	4	5	6	7	8	9
10	11	12	13	14	15	16
17	18	19	20	21	22	23
24	25	26	27	28		

March 3003

Mo	Tu	We	Th	Fr	Sa	Su
					1	2
3	4	5	6	7	8	9
10	11	12	13	14	15	16
17	18	19	20	21	22	23
24	25	26	27	28	29	30
31						

16 January 6th
17 February 18th
18 March 9th
19 April 4th
20 May 1st
21 May 12th

E Write each date in figures.

22 12.9.08

22 12th September 2008
23 5th March 2005
24 21st October 2009
25 8th August 2007
26 1st January 2010

Challenge

Work out which day of the week it will be in Navros in 3003 on:
a your birthday *b* Christmas Day *c* Bonfire Night (November 5th).

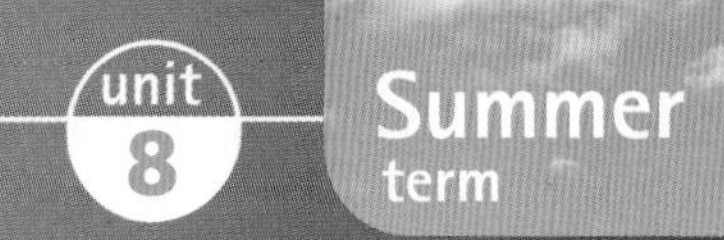

Costalot Coach Timetable

	coach 1	coach 2	coach 3
Orton	7:30 a.m.	9:20 a.m.	11:10 a.m.
Walsey	7:55 a.m.	9:40 a.m.	11:30 a.m.
Minton	8:15 a.m.	10:00 a.m.	11:50 a.m.
Kenton	8:40 a.m.	10:25 a.m.	12:15 p.m.
Garfield	9:15 a.m.	10:50 a.m.	12:50 p.m.
Thorpe	9:25 a.m.	11:00 a.m.	1:00 p.m.

A Use the coach timetable to answer these questions.

1 1 hour 50 minutes

1. How long between the starting times of each coach?
2. How long does coach 1 take to complete the whole journey?
3. You arrive at Orton at 9:30 a.m. to travel to Kenton. At what time will you arrive in Kenton?
4. You arrive at Minton at 8:50 a.m. How long must you wait for a coach?
5. Coach 2 is running 20 minutes late. What time does it arrive in Garfield?
6. You just miss coach 1 at Walsey. How long must you wait for the next coach?
7. Which is the quickest coach journey?
8. Which is the slowest coach journey?
9. Explain why you think one journey takes longer than another.
10. Which coach would you take to visit a friend in Garfield at 11:00 a.m.?

B Work out the time taken for each journey.

11. coach 1: Walsey to Kenton
12. coach 2: Orton to Kenton
13. coach 3: Walsey to Thorpe
14. coach 1: Minton to Thorpe
15. coach 2: Orton to Garfield

C Look at the coach timetable.

16. Rewrite all the times using the 24-hour clock.

Challenge

The boss of Costalot Coaches wants to start a new coach 4. It will set off from Orton at 5:25 p.m. He predicts coach 4 will take the same amount of time as coach 1 to travel between the towns. Write a timetable for coach 4.

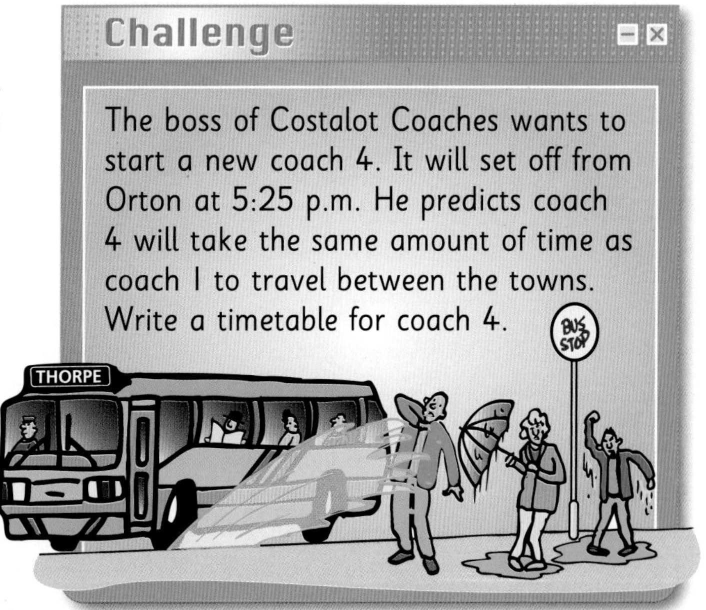

A Complete these.

1 $5 \times 5 =$ ✹
2 $6 \times 10 =$ ●
3 $3 \times 100 =$ ✸
4 double 50 = ●
5 $20 \times 4 =$ ▲
6 $30 \times$ ● $= 300$
7 ■ $\times 10 = 500$
8 $90 \div 10 =$ ●
9 $70 \times 10 =$ ●
10 $\frac{1}{2}$ of 600 = ■
11 $20 \times$ ★ $= 100$

B Multiply each number by 10.

12 16
13 43
14 28
15 123
16 426
17 100
18 95
19 209

C Write the number that will come out of the function machine for each input number.

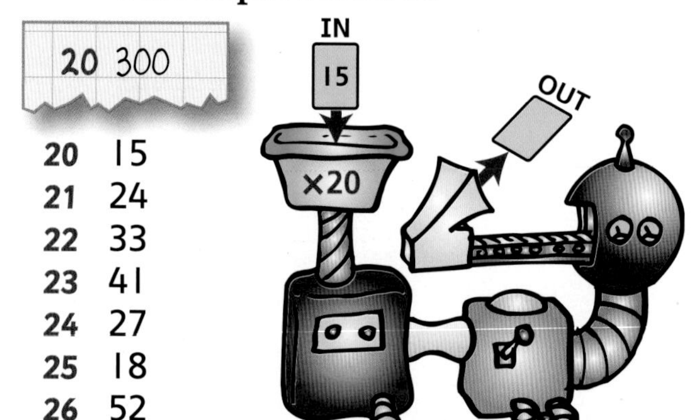

20 15
21 24
22 33
23 41
24 27
25 18
26 52

D Write the number that will come out of the function machine for each input number.

OUT
IN
32
×5

27 32
28 26
29 24
30 40
31 48
32 34
33 19

E Divide each number by 10.

34 4360
35 490
36 640
37 1240
38 5680
39 2030

F Write three facts linked to each number sentence.

40 $3 \times 11 = 33$
41 $5 \times 12 = 60$
42 $12 \times 4 = 48$
43 $9 \times 8 = 72$
44 $20 \times 6 = 120$

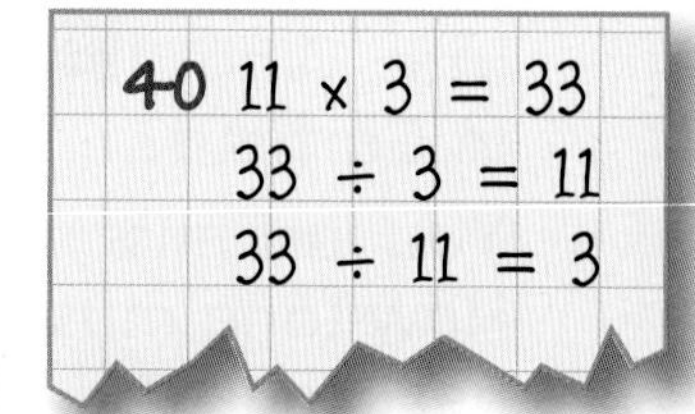

Challenge

Copy and complete each multiplication and division chain.

a 15 → ×10 → ×2 → ÷100 → ○
b 20 → ×10 → ×5 → ÷10 → ○
c 44 → ×20 → ÷10 → ×100 → ○

880

Make two multiplication and division chains of your own:

d start with 100
e end with 1000.

Can you understand the relationship between multiplication and division?

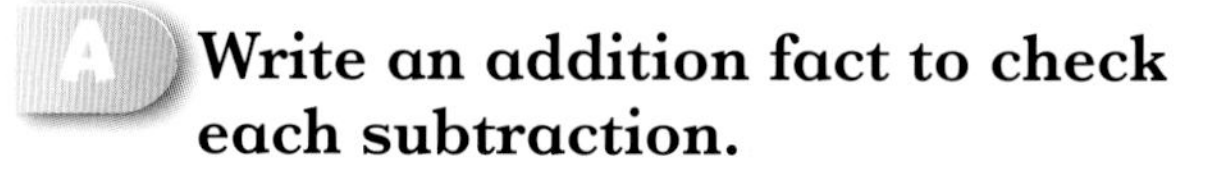

A Write an addition fact to check each subtraction.

1 42 − 17 = 25
2 37 − 19 = 18
3 64 − 28 = 36
4 101 − 52 = 49
5 60 − 19 = 41

1 25 + 17 = 42

B Write a multiplication fact to check each division.

6 36 ÷ 4 = 9
7 27 ÷ 3 = 9
8 55 ÷ 5 = 11
9 32 ÷ 4 = 8
10 36 ÷ 3 = 12
11 202 ÷ 2 = 101

C Use facts from the tool box to help you answer these.

12 19 × 3
13 51 ÷ 3
14 111 ÷ 3
15 16 × 6
16 23 × 5
17 57 ÷ 19
18 17 × 3
19 111 ÷ 37

3 × 17 = 51
57 ÷ 3 = 19
96 ÷ 16 = 6
37 × 3 = 111
115 ÷ 5 = 23
TOOLS

D Answer these. Write the remainder for each.

20 22 ÷ 5 = 4 r 2

20 22 ÷ 5
21 221 ÷ 2
22 112 ÷ 10
23 47 ÷ 4
24 32 ÷ 3
25 123 ÷ 5

E Solve these problems.

26 Shelf units need 5 brackets. How many shelves can be put up with 28 brackets?

27 A class of 30 children is divided into teams of 7. How many teams can be made?

28 Each table in a restaurant can seat 6 people. How many tables are needed for 33 people?

29 How many bicycles can be made with 63 wheels?

30 Helicopters can carry 9 passengers. How many helicopters are needed for 86 passengers?

Challenge

Follow the trail of correct remainders to find out which type of transport takes off from Beefro Airport.

START
27 ÷ 4 1 3
36 ÷ 10 6 3
101 ÷ 2 1 11
39 ÷ 5 4 3
92 ÷ 10 2 9
49 ÷ 5 4 9
68 ÷ 10 8 6
43 ÷ 4 1
601 ÷ 10 1 6
38 ÷ 3 2 1
68 ÷ 5 1 3
19 ÷ 4 2 3
Helicopter
Concorde
Glider
Jet
Microlight
Balloon

A Solve these problems so Kian can have a happy birthday party.

Please come to
Kian's Birthday Party
on Tuesday at 4 o'clock
at Jungle Gym FunZone

There will be 12 children at the party.

1 Each child is given 3 sweets. How many sweets are needed?

2 Each child receives a present costing £1·99. What is the total cost of the presents?

3 For tea the children have chicken nuggets and chips for £1·95, an ice-cream for 77p, a banana for 23p and a drink for 35p. What does each child's meal cost?

4 There are 50 balloons at the party. Each child bursts 3. How many balloons are left?

5 Nicky and Andrew have £10 to buy presents. Nicky spends £3·62. Andrew spends £2·99. How much change is left?

6 3 children each bring Kian a pack of 48 trading cards. How many trading cards does Kian get?

7 Ayeesha isn't hungry. She only eats 2 items from question 3. Her meal costs £1·12. What does Ayeesha eat?

8 5 children come to the party by bus. The total fare is £2·95. What is the fare for 1 child?

B Copy and complete.

9 (7 × 7) = (5 × 7) + (2 × 7)

9 (7 × 7) = (5 × 7) + (● × 7)
10 (12 × 9) = (● × 9) + (2 × 9)
11 (24 × 6) = (10 × 6) + (10 × 6) + (● × 6)
12 (14 × 8) = (10 × 8) + (4 × ■)
13 (17 × 5) = (10 × 5) + (▲ × 5)

C Answer these.

14	23 × 4	17	28 × 4	20	53 × 8
15	33 × 6	18	46 × 6	21	62 × 9
16	41 × 5	19	29 × 7	22	37 × 12

Challenge

Play with a partner. Take turns to roll two dice and make the largest 2-digit number you can. Roll one dice again. Multiply your 2-digit number by the dice number. The player with the higher answer scores a point.

The first player to reach 5 points wins.

53 × 4 = 212

Can you use different strategies for solving problems?

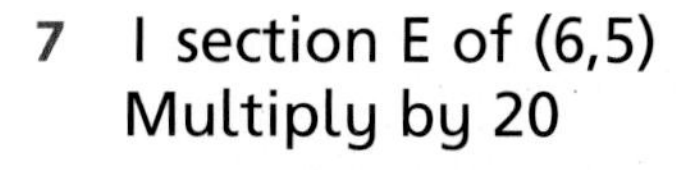

Follow each clue to find a number. Carry out the calculation on the number you find.

1. 2 sections SE of (6,4)
   Multiply by 13
2. 4 sections N of (3,0)
   Divide by 5
3. Red hand 60° clockwise
   Double the number
4. 1 section NE of (1,3)
   Halve the number
5. Blue hand 180° anti-clockwise
   Divide by 10
6. 7 sections NE of (0,0)
   Multiply by 5
7. 1 section E of (6,5)
   Multiply by 20
8. Red hand 120° anti-clockwise
   Multiply by 3
9. 3 sections E of (1,1)
   Divide by 5
10. 2 sections SW of (8,4)
    Multiply by 8
11. Red hand 150° clockwise
    Multiply by 6
12. Blue hand 30° clockwise
    Multiply by 4
13. 7 sections SW of (8,8)
    Subtract 46
14. 3 sections SW of (4,7)
    Divide by 10
15. Blue hand 120° anti-clockwise
    Multiply by 12

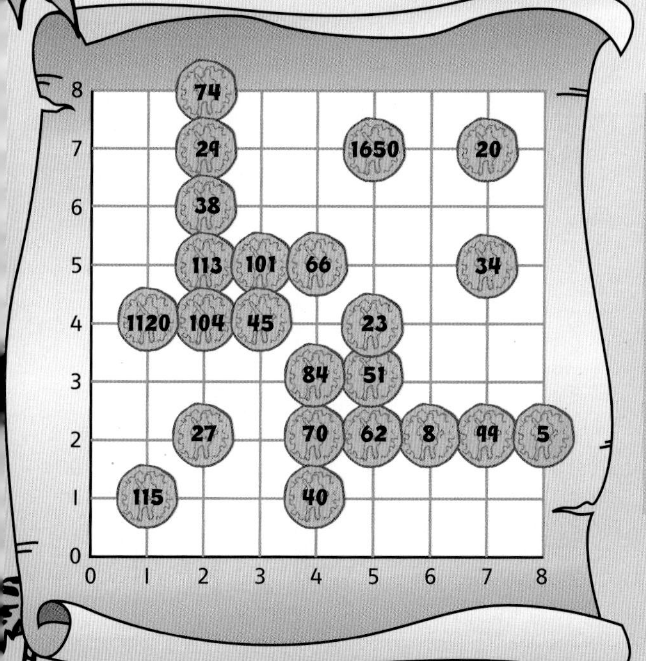

Add together all your answers. Use a calculator to help.

The thousands digit is the first number in a pair of co-ordinates.

The units digit is the second number in a pair of co-ordinates.

The co-ordinates tell you where on the map to find the treasure. How many gold coins are buried there?

A Copy and complete. Write the colour of the circle that contains each answer.

1 6 × 10
2 8 × 2
3 7 × 5
4 4 × 10
5 9 × 2
6 5 × 5
7 3 × 5
8 7 × 2
9 9 × 10
10 9 × 5
11 8 × 5

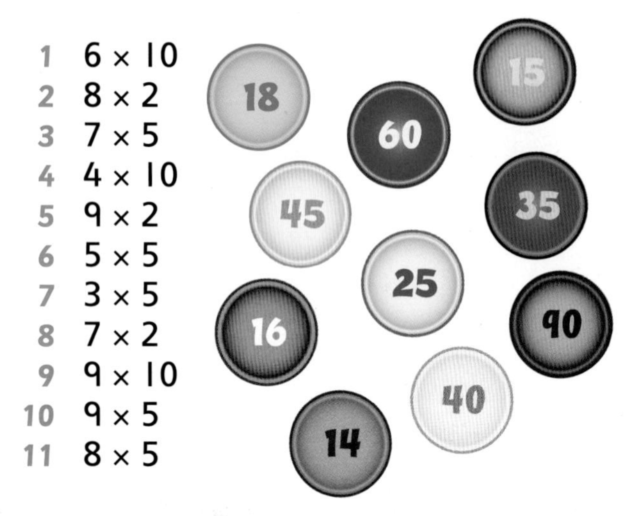

B Write how much is needed for 10 people to each have these amounts.

12 £82
13 £23
14 £65
15 £132
16 £402
17 £80
18 £124
19 £365

C Share these animals between 10 farmers. Write how many each.

20 110 cows
21 130 sheep
22 420 goats
23 90 horses
24 1350 turkeys
25 3260 chickens

D Multiply each number by 100.

26 42
27 61
28 78
29 361
30 415
31 836

E In a sports stadium seats are arranged in blocks of 100. Find how many blocks are filled by each crowd.

32 60 blocks

32 6000 people
33 9000 people
34 6500 people
35 3700 people
36 8900 people
37 7200 people

Challenge

Play with a partner. Use two sets of 12 counters and two dice.
Take turns to roll one or both dice.
Count around sections on the Times Track to match the total dice number.
Multiply the total dice number by 10 or 100 as shown in the section you land on. If your answer appears on the grid cover it with a counter. The first player with a line of four counters wins.

110	500	40	100	80
600	10	700	1100	30
70	1200	100	200	120
1000	400	20	800	60
50	300	100	90	900

Can you multiply a 2-or 3-digit number by 10 or 100 and divide a 4-digit multiple of 1000 by 10 or 100?

A Use the grid method to answer these.

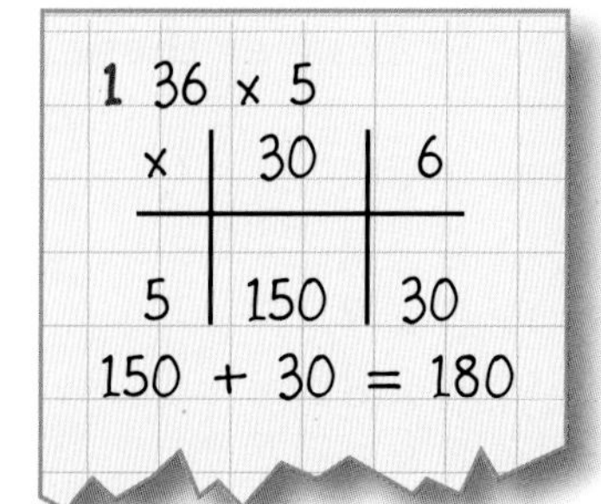

1 36 × 5
2 37 × 4
3 29 × 3
4 46 × 4
5 53 × 6
6 28 × 7
7 38 × 9
8 54 × 8

B Answer these. Show your working. Write the remainder if there is one.

9 37 ÷ 5
10 34 ÷ 3
11 28 ÷ 5
12 46 ÷ 4
13 107 ÷ 10
14 88 ÷ 4
15 74 ÷ 6
16 82 ÷ 7
17 59 ÷ 6

C Write the shape that contains each number.

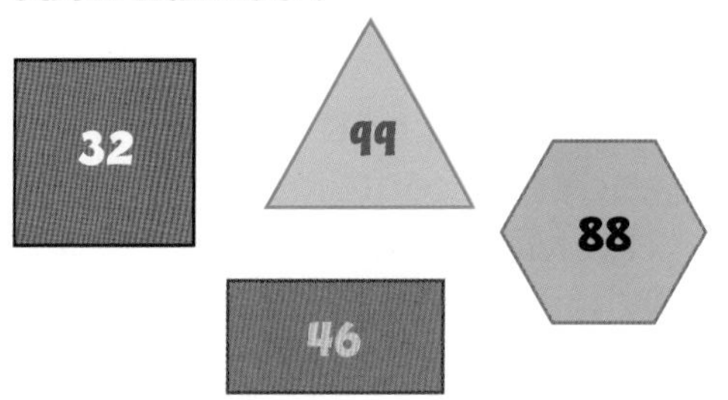

18 has a remainder of 2 when divided by 3

18 square

19 has a remainder of 2 when divided by 4
20 has a remainder of 4 when divided by 5
21 has a remainder of 8 when divided by 10
22 has a remainder of 1 when divided by 2
23 has a remainder of 1 when divided by 3 or 5

Challenge

The letter and numbers are the key to a code.

D	Y	G	S	O	A	X	F
21	64	42	32	30	18	45	48

R	L	P	U	T	E	C	V
35	80	63	28	60	27	81	72

W	K	J	N	B	M	H	I
70	49	36	56	42	54	40	24

a Decode this question:

(10 × 4) (10 × 3) (10 × 7)
(9 × 6) (9 × 2) (7 × 8) (8 × 8)
(3 × 7) (6 × 3) (8 × 8) (8 × 4)
(3 × 8) (8 × 7) (6 × 3)
(10 × 7) (3 × 9) (9 × 3) (7 × 7)?

b Give your answer in code using number facts from the 8 or 9 times tables.

c Write a coded question for a partner to answer.

A Solve these problems. Show your working.

1 How many bottles in 26 boxes if each box contains 6 bottles?

2 The Millennium Wheel takes 34 minutes for each rotation. How many minutes for 6 rotations?

3 There are 16 people in each pod on the wheel. How many people altogether in 5 pods?

4 There are 5 balloons in a pack. How many packs do you need to buy to have 63 balloons?

5 A fairground ride costs £4 per person. What is the cost for a group of 13 people?

6 How many each and how many left over when 5 children share 47 fair ride vouchers equally among them?

7 The Ghost Train takes 4 people in each carriage. How many carriages are needed for 66 people?

8 24 bags of candyfloss can be made with 1 kg of sugar. How many bags will 8 kg of sugar make?

9 There are 8 sausages in a pack. How many packs are needed to make 26 hot dogs with 2 sausages each?

B Answer these.

10 7×8
11 6×9
12 5×7
13 4×8
14 6×4
15 5×5
16 7×9
17 8×8
18 4×9
19 9×8
20 8×7
21 7×7
22 9×6
23 9×9
24 6×5
25 6×8

Challenge

Use the six digits in each set to make three 2-digit numbers. Use each digit once only. The 2-digit numbers must be the multiples shown.

a multiple of 3
multiple of 4
multiple of 5
4 5 6 7 8 9

b multiple of 2
multiple of 3
multiple of 4
3 4 5 6 7 8

c multiple of 4
multiple of 5
multiple of 6
0 1 2 3 4 5

d multiple of 4
multiple of 5
multiple of 6
1 2 3 4 5 6

A Write the fraction of each shape that is shaded.

1 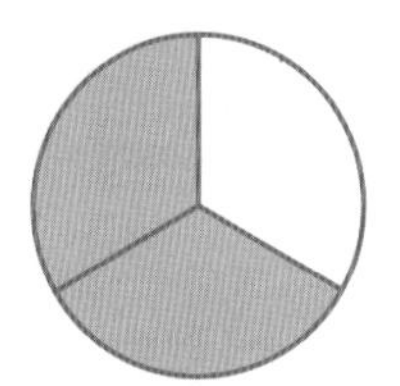

2 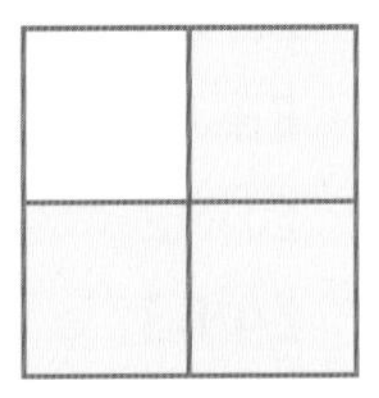

3 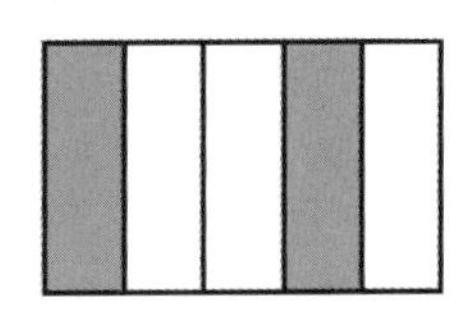

4 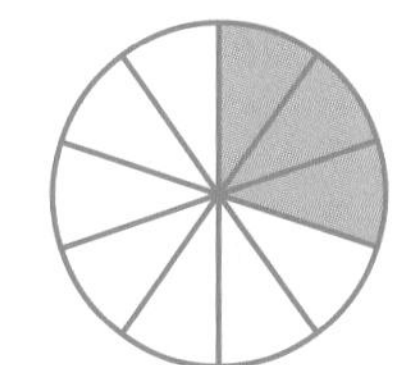

5 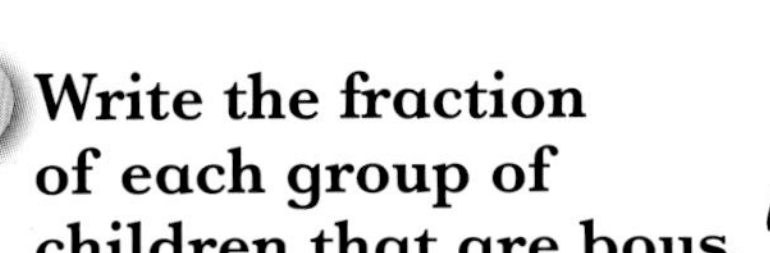

B Write the fraction of each group of children that are boys.

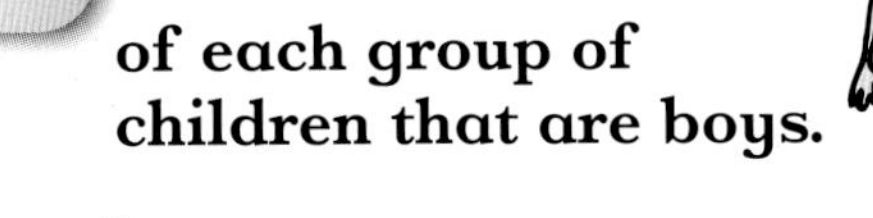

6 $\frac{7}{10}$ are girls

7 $\frac{3}{4}$ are girls

8 $\frac{3}{10}$ are girls

9 $\frac{5}{8}$ are girls

10 $\frac{1}{5}$ are girls

11 $\frac{1}{3}$ are girls

12 $\frac{7}{8}$ are girls

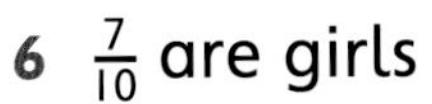

C Copy and complete.

13 $\frac{9}{10}$ + ✹ = 1

14 $\frac{1}{3}$ + ◖ = 1

15 $\frac{3}{4}$ + ✳ = 1

16 $\frac{4}{5}$ + ● = 1

17 $\frac{7}{12}$ + ▲ = 1

18 $\frac{3}{10}$ + ● = 1

19 $\frac{9}{16}$ + ■ = 1

D Write what fraction of marbles are red in each set.

20 $\frac{3}{10}$

20 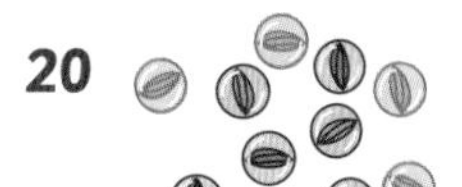

22 

21 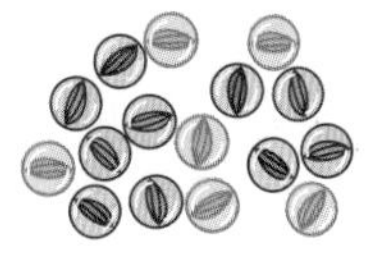

23 

24 

E Copy and complete.

25 $\frac{1}{2} = \frac{◖}{6}$

26 $\frac{1}{2} = \frac{✳}{8}$

27 $\frac{3}{4} = \frac{6}{●}$

28 $\frac{2}{5} = \frac{4}{✹}$

29 $\frac{1}{2} = \frac{5}{★}$

30 $\frac{2}{▲} = \frac{1}{4}$

31 $\frac{4}{5} = \frac{●}{10}$

32 $\frac{6}{12} = \frac{■}{2}$

33 $\frac{4}{12} = \frac{1}{■}$

Challenge

Copy each diagram onto squared paper. For each diagram shade the fraction of the shape shown. Write the fraction that is not shaded for each.

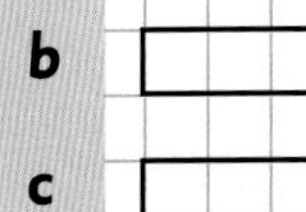

a $\frac{3}{5}$

b $\frac{1}{2}$

c $\frac{2}{7}$

d $\frac{5}{8}$

e $\frac{2}{3}$

Do you know the equivalence between halves, quarters and eighths, tenths and fifths, and thirds and sixths?

A Write as decimal fractions.

1 $\frac{3}{10}$ 2 $\frac{1}{5}$ 3 $\frac{1}{10}$ 4 $\frac{3}{5}$ 5 $\frac{7}{10}$ 6 $\frac{9}{10}$ 7 $\frac{4}{5}$ 8 $\frac{1}{2}$

1 0·3

B Write as fractions.

9 0·3 10 0·6 11 0·9 12 0·1 13 0·7 14 0·4 15 0·8 16 0·5

C Write as decimals.

17 $3\frac{2}{5}$ 18 $2\frac{4}{10}$ 19 $1\frac{3}{5}$ 20 $4\frac{1}{2}$ 21 $3\frac{9}{10}$ 22 $2\frac{7}{10}$

17 3·4

D Copy and complete the table.

a 0 – 2

b 3 – 5

c 6 – 8

	line	flag colour	decimal number
23	a	red	0·1
24	b	red	
25	c	yellow	
26	a	purple	
27	b	blue	
28	c	blue	
29	a	pink	
30	b	purple	
31	c	pink	

E Write each amount in pence.

32 £2·07 33 £3·15 34 £1·92 35 £4·63 36 £7·02 37 £10·23 38 £9·09

32 207p

F Write each length in metres.

39 64 cm 40 38 cm 41 85 cm 42 196 cm 43 223 cm 44 390 cm 45 401 cm

39 0·64 m

Challenge

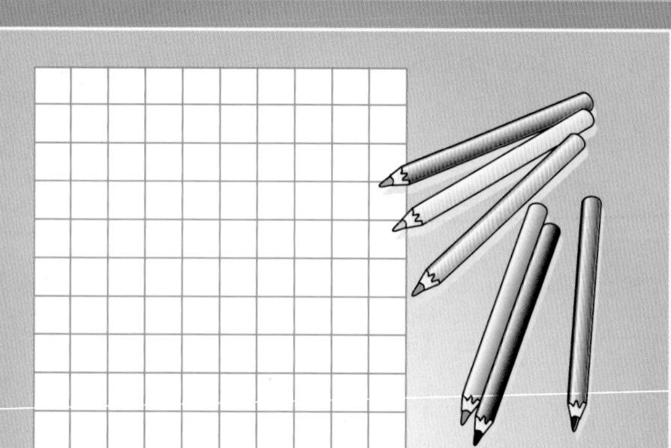

Copy the square onto squared paper. Make an interesting pattern by colouring the square to follow these rules:

a $\frac{1}{5}$ of the squares are red
b 0·25 of the squares are blue
c $\frac{3}{10}$ of the squares are orange
d 0·08 of the squares are yellow
e 0·1 of the squares are black
f 0·07 of the squares are pink.

Do you know what each digit in a decimal fraction represents?

A Copy and complete.

1 1 in every 3 trucks is red

I in every ✸ trucks is red

I in every ● trucks is purple

I in every ● trucks is yellow

I in every ● trucks is blue

2 in every ★ trucks is orange

2 in every ▲ trucks is pink

B Write what fraction of the trucks in each train are these colours.

7 purple trucks in train 2
8 orange trucks in train 5
9 yellow trucks in train 3
10 blue trucks in train 4
11 pink trucks in train 6

C Draw strings of beads for each pattern.

12 I in every 3 is yellow
13 I in every 4 is red
14 3 in every 8 are purple
15 4 in every 12 are blue

D Write as fractions.

16 0·73
17 0·64
18 0·94
19 0·38
20 0·5
21 0·25
22 0·75

E Copy and complete.

23 $\frac{30}{100} = \frac{▲}{10}$

24 $\frac{40}{100} = \frac{4}{●}$

25 $\frac{90}{100} = \frac{●}{10}$

26 $\frac{20}{100} = \frac{●}{10}$

27 $\frac{1}{2} = \frac{✸}{100}$

28 $\frac{60}{100} = \frac{★}{10}$

29 $\frac{1}{4} = \frac{■}{100}$

30 $\frac{3}{4} = \frac{75}{●}$

Challenge

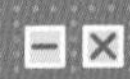

Draw 12 garden gnomes.
Make sure that:

a I in every 3 has a black hat
b 2 in every 12 have red buttons
c 6 in every 12 have blue trousers
d I in every 4 has a yellow shirt
e 9 in every 12 have green shoes.

Can you describe patterns using the language 'in every'?

Fraction bingo

You need:

- two sets of cards numbered 1 to 7
- two sets of nine counters
- seven cards labelled $\frac{1}{2}, \frac{1}{4}, \frac{1}{3}, \frac{2}{3}, \frac{3}{4}, \frac{5}{12}, \frac{7}{12}$
- a partner

Each take a set of counters and choose a game board.

Shuffle the fraction cards and deal one to each player. This fraction tells you the number of squares you need to cover on your board to win,
e.g. $\frac{2}{3}$ of 12 squares = 8 squares.

Each take one set of number cards and shuffle them. Both players turn over the top card in their set of cards. Multiply the numbers together. If the answer appears on your game board, cover it with a counter.

When all the number cards have been used, shuffle them and play again.

The winner is the first player to cover the number of squares shown by their fraction card.

35	20	6	30
14	5	49	12
21	15	1	10

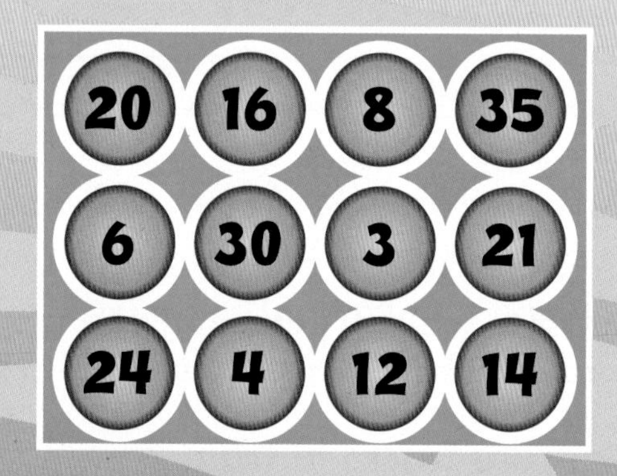

A Complete these. Show your working.

1 63 + 28
2 46 + 33
3 52 + 29
4 36 + 47
5 28 + 34
6 36 + 56
7 19 + 73
8 16 + 49
9 27 + 68

B Write each missing number.

10 52 + □ = 100
11 45 + □ = 100
12 181 + □ = 200
13 391 + ▲ = 400
14 666 + □ = 700
15 991 + □ = 1000
16 926 + □ = 1000

C Answer these.

17 62 − 28
18 54 − 26
19 82 − 37
20 28 − 19
21 74 − 37
22 91 − 19
23 84 − 38

D Use the information in the box to answer these.

24 101 − 83
25 163 − 95
26 693 + 198
27 424 − 58
28 503 − 76
29 891 − 693
30 592 + 479

83 + 18 = 101
1071 − 479 = 592
58 + 366 = 424
68 + 95 = 163
76 + 427 = 503
891 − 198 = 693

E Solve these money problems.

£1026 £8 £700 £54

31 How much more money than Ivor does Miss Moneybags have?
32 Miss Moneybags gives Arthur £30. How much has she left?
33 How much more money than Arthur does Ivor have?
34 Do all four people together have enough money to buy a boat for £1800?
35 How much more money than Penny does Ivor have?

Challenge

Use a 100 square to help you.
This grid is part of a 100 square. Add together:

46	47	48
56	57	58
66	67	68

a the numbers in the opposite corners
b the opposite middle numbers
c the numbers along each diagonal.
What do you find?
Try this with other 3 × 3 grids and larger grids from the 100 square.

A Copy and complete.

1. 600 − 4 = ✹
2. 800 − 5 = ◖
3. 300 − ✳ = 289
4. 625 − ● = 617
5. 293 + ▲ = 304
6. 900 − 8 = ◖
7. 1001 − ✿ = 998
8. 700 − ■ = 693
9. 492 + ★ = 500
10. 5000 − ● = 4900
11. 3000 − 9 = ■

B Add each pair of numbers. Show all your working.

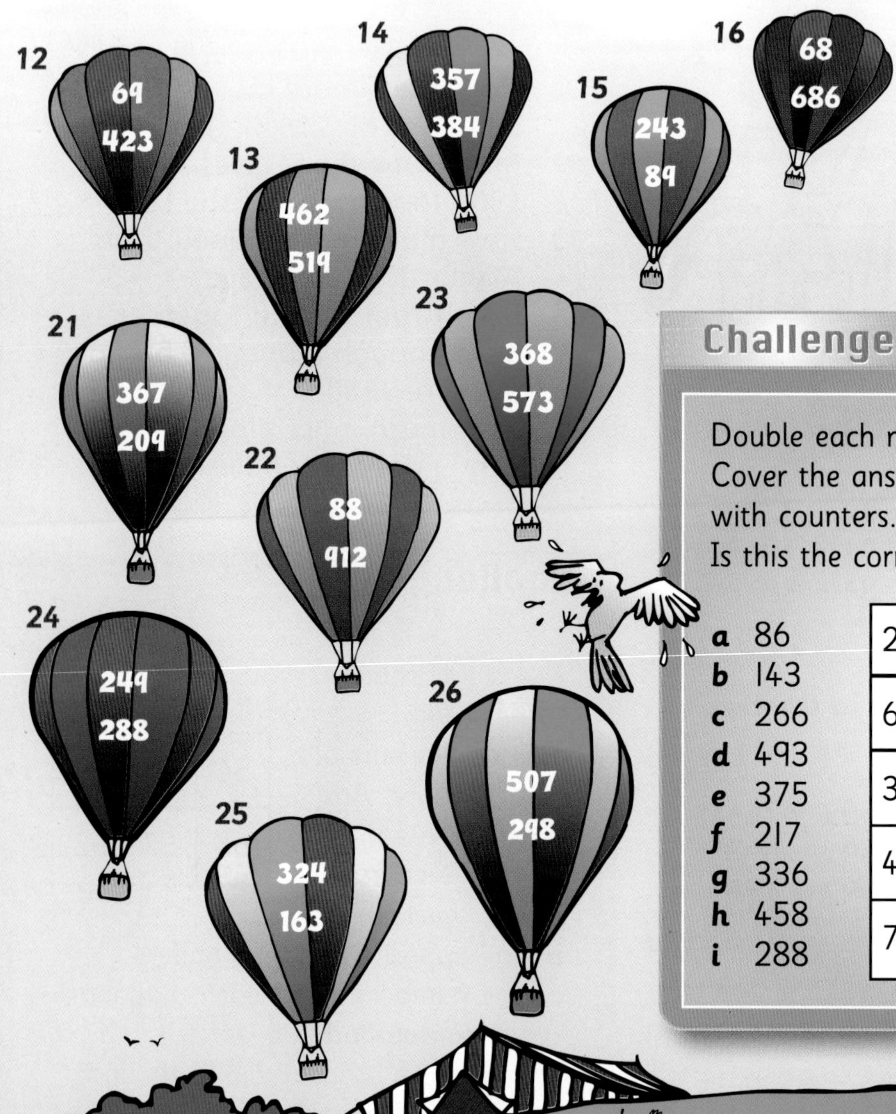

12. 69, 423
13. 462, 519
14. 357, 384
15. 243, 89
16. 68, 686
17. 324, 466
18. 810, 99
19. 247, 583
20. 624, 168
21. 367, 209
22. 88, 912
23. 368, 573
24. 249, 288
25. 324, 163
26. 507, 298

Challenge

Double each number.
Cover the answers on the grid with counters.
Is this the correct mark for your work?

a 86
b 143
c 266
d 493
e 375
f 217
g 336
h 458
i 288

286	344	650	572	916
634	986	616	172	534
386	675	672	712	786
432	576	632	532	567
750	816	272	850	434

Can you use an appropriate strategy to add together any two numbers less than 1000?

A Use a written method to answer these. Show all your working.

1 384 − 158
2 728 − 261
3 435 − 167
4 506 − 125
5 463 − 295
6 804 − 196
7 774 − 187
8 402 − 124
9 632 − 478
10 911 − 119
11 620 − 176

B Find the missing digits for each calculation.

12
```
  2 7 1
+ 3 ✸ 2
  6 1 3
```

13
```
  3 0 5
+ 1 6 ●
  4 7 2
```

14
```
  2 ■ 4
+ 1 6 6
  4 6 0
```

15
```
  6 3 1
− 2 2 ●
  4 0 3
```

16
```
  2 4 6
−   8 ▲
  1 5 9
```

17
```
  5 ● 3
− 1 4 ■
  4 4 7
```

18 327 − ✸▲6 = 81

19 4✸■ + 163 = 592

C Copy and complete.

20 4632 + 1653
21 1751 + 864
22 3862 + 1785
23 6237 + 1604
24 4976 + 1638
25 2468 + 1357
26 8762 + 1098

Challenge

5

7 8

Use these digits once each to make a 2-digit and a 3-digit number. Add your two numbers together and write the total. Repeat, making different 2- and 3-digit numbers.

a What is the largest total you can make?

b What is the smallest total you can make?

Find the largest and smallest totals you can make using five other digits in this way.

A Use the bar chart to answer each question.

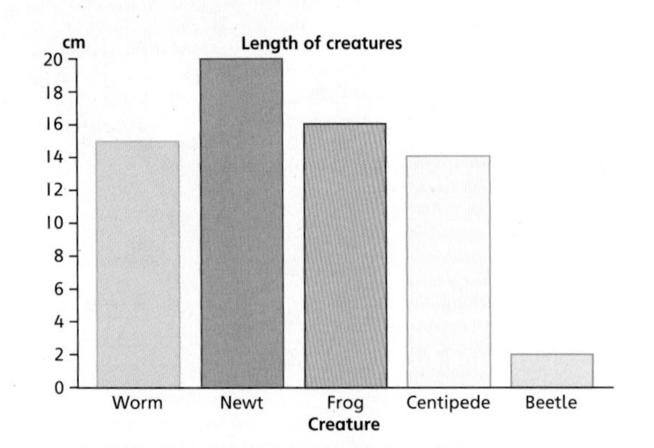

1 How long is the newt?
2 How long in the worm?
3 How long is the frog?
4 How long is the beetle?
5 How much longer is the worm than the beetle?
6 How much shorter is the centipede than the frog?
7 If the creatures line up end to end, how far will they stretch?

B Copy and complete the Venn diagram. Answer the questions.

8 Enter the numbers 1 to 30 on your diagram.
9 How many numbers less than 30 are multiples of 4?
10 How many numbers less than 20 are not multiples of 4?
11 How many numbers less than 30 are not multiples of 4?
12 How many numbers less than 20 are multiples of 4?

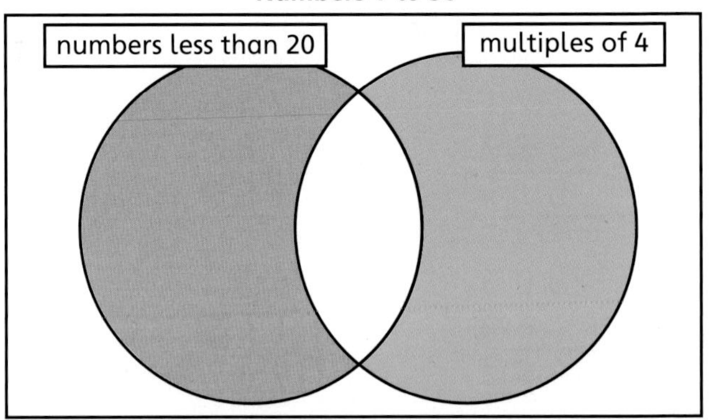

Challenge

Play with a partner. Copy the Venn diagram. Take turns to roll two dice and multiply your numbers together. Write the answer in the correct section of the diagram. Score points as follows:

a in the pink – 5 points
b in either circle but not in the pink – 3 points
c not in either circle – lose 1 point.

If your answer is already on the diagram, miss a turn. The first player to reach 25 points wins.

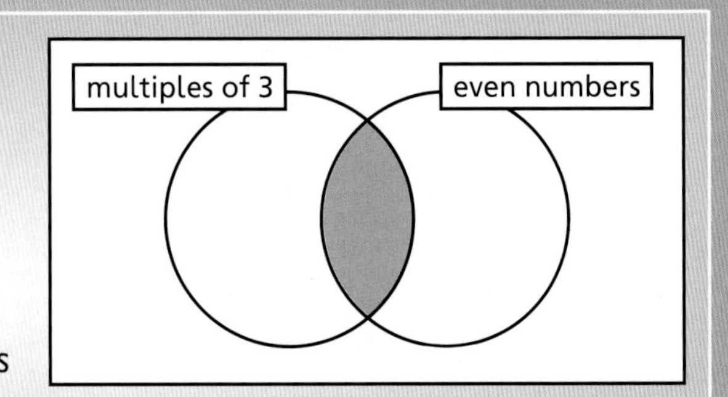

Can you solve problems using a Venn diagram?

A Copy and complete the Carroll diagram. Use the numbers on the motorbikes.

1

	red bike	not red bike
odd number	3	
not odd number		

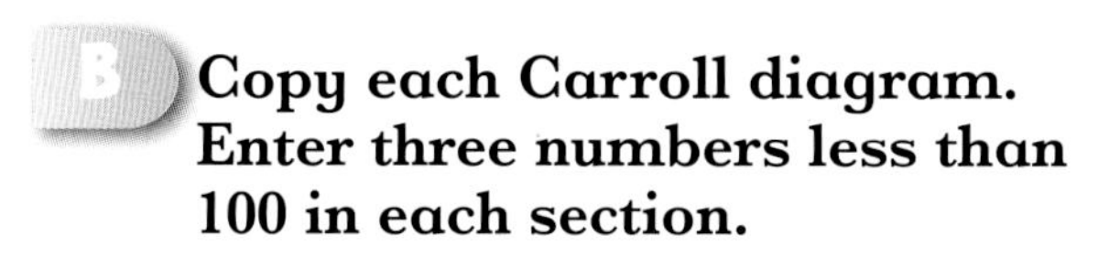

B Copy each Carroll diagram. Enter three numbers less than 100 in each section.

2

	even	not even
multiple of 5	20 60 70	
not a multiple of 5		

3

	multiple of 10	not a multiple of 10
multiple of 3		
not a multiple of 3		

4

	number greater than 60	numbers not greater than 60
multiple of 10		
not a multiple of 10		

5

	numbers having 5 as digit	numbers not having 5 as digit
numbers greater than 55		
numbers not greater than 55		

6

	numbers with units digit greater than tens digit	numbers with units digit not greater than tens digit
numbers less than 50		
numbers not less than 50		

Challenge

Copy and complete this Carroll diagram by entering the initials of the children in your class.

	children older than me	children not older than me
children taller than me		
children not taller than me		

Can you use a Carroll diagram to sort information?

A **Use the pictogram to find how many balloons of each colour were sold.**

1 red balloons
2 green balloons
3 yellow balloons
4 blue balloons
5 all coloured balloons

1 20

Number of balloons sold at the fair

Red	🎈🎈🎈🎈
Yellow	🎈🎈
Green	🎈🎈🎈🎈🎈🎈🎈
Blue	🎈🎈🎈🎈🎈

🎈 represents 5 balloons

B **Use the pictogram to answer these.**

6 How many pies were thrown on Monday?
7 How many pies were thrown on Tuesday?
8 How many pies were thrown on Wednesday?
9 What do you think happened on Wednesday?
10 How many pies were thrown on Thursday?
11 How many pies were thrown on Friday?

Number of custard pies thrown by the clowns

Monday	○○○○○
Tuesday	○○○ ◖
Wednesday	
Thursday	○○○○ ◖
Friday	○○○○○ ◺

○ represents 4 custard pies

C **Use this symbol ▭ to represent 20 tickets. Draw a pictogram to display the information in the table.**

12

Performance Day

	Tuesday	Wednesday	Thursday	Friday	Saturday
Tickets sold for concert	120	80	90	105	75

Challenge

Ask the children in your class to choose their favourite pizza topping from the menu. Design a pictogram to display the information.

Menu

Sausage, bacon & mushroom
Vegetarian
Ham & pineapple
Spicy chicken, pepper & onion
Pepperoni & mushroom
Meat feast
Cheese feast

PIZZA

Can you represent information in a pictogram?

A Write how long each journey takes.

	Start time	Finish time
1	7:15 a.m.	8:05 a.m.
2	7:32 a.m.	9:00 a.m.
3	10:15 a.m.	11:50 a.m.
4	10:34 a.m.	12:16 p.m.
5	8:14 p.m.	10:04 p.m.
6	3:47 p.m.	6:12 p.m.

B Copy and complete. Write the remainder.

7 $39 \div 2$
8 $40 \div 3$
9 $58 \div 5$
10 $111 \div 10$
11 $43 \div 4$
12 $101 \div 2$

C Copy and complete.

13 8×5
14 7×2
15 9×3
16 10×10
17 4×9
18 6×5

D Use the grid method to answer these.

19 38×4
20 54×7
21 48×6
22 56×8

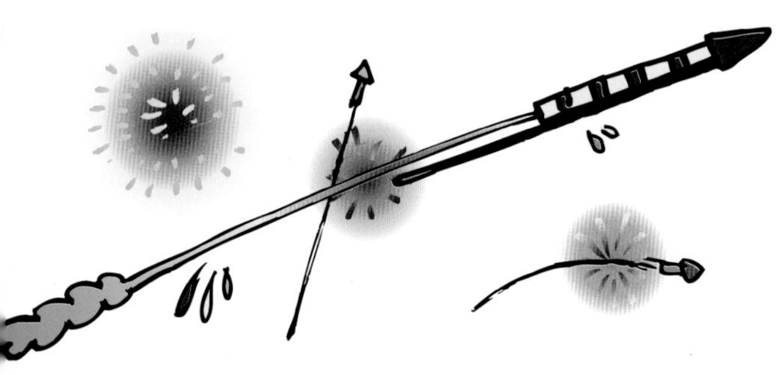

E Write as decimal fractions.

23 $\frac{3}{10}$
24 $\frac{1}{5}$
25 $\frac{7}{10}$
26 $\frac{1}{2}$
27 $\frac{3}{5}$
28 $\frac{1}{4}$

F Write as fractions.

29 0.7
30 0.4
31 0.5
32 0.75

G Complete these.

33 $327 + 48$
34 $467 + 149$
35 $632 + 176$
36 $339 + 285$

H Copy and complete the Venn diagram.

Numbers 1 to 30

37

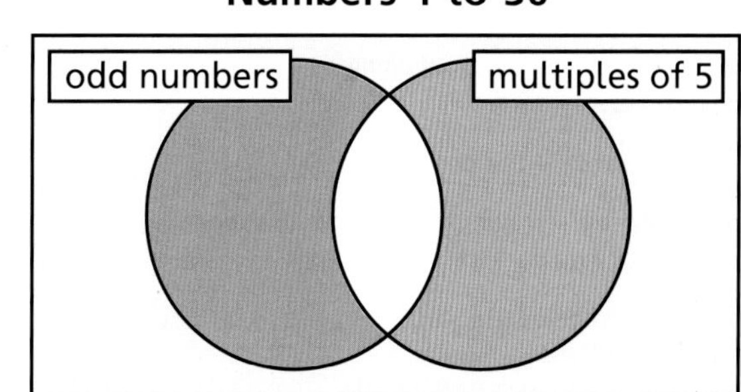

I Copy and complete the Carroll diagram.

Numbers 1 to 30

38

	odd number	not odd number
multiple of 3		
not a multiple of 3		

Glossary

24-hour clock – measures time starting from 0 to 24 hours, e.g. 05:30 is 5:30 a.m. and 17:30 is 5:30 p.m.

a.m. – short for 'ante meridiem' meaning 'before noon', used to show times between 12 midnight and 12 noon

anti-clockwise – turns in the direction opposite to the hands on a clock

approximate answer – a rough answer or estimate near to the exact answer

area – the amount of surface a shape covers, measured in square units such as cm^2 and m^2

array – a regular arrangement of objects in rows and columns

axis (axes) – graphs have two axes, one horizontal and one vertical

capacity – the amount a container holds, measured in l and ml

Carroll diagram – used for sorting things into groups, e.g. red and not red, cubes and not cubes

clockwise – turns in the same direction as hands on a clock

consecutive – things that come one after another in a regular order, e.g. 11, 13, 15 are consecutive odd numbers

co-ordinates – pairs of numbers that show the position of a point on a graph, read across first (horizontal co-ordinate), then up (vertical co-ordinate)

data – information about something in words, numbers or pictures

denominator – the bottom number of a fraction

divisible by – a number is divisible by another number if it can be divided exactly with no remainder, e.g. 49 is divisible by 7

equilateral triangle – a triangle with all sides the same length and all angles the same

equivalent fractions – fractions with the same value, e.g. $\frac{4}{6} = \frac{2}{3}$

estimate – using information you have to guess an answer without measuring or doing a difficult calculation

horizontal – a level or flat line parallel to the horizon or ground, a line parallel to the bottom edge when represented on paper

inverse – the opposite, addition and subtraction are inverse operations

irregular polygon – sides or angles are not equal

isosceles triangle – a triangle with two sides the same length and two angles the same

line symmetry – a shape has line symmetry if it can be folded so one half covers the other exactly

mass – the amount of matter in an object, measured in g or kg (sometimes people use weight to mean mass)

mid-point – halfway between two points

mixed number – a number containing a whole number and a fraction part, e.g. $2\frac{1}{4}$

multiple – a number that is exactly divisible by another, e.g. numbers in the times tables, 5, 10, 15, 20, 25, 30 are all multiples of 5

negative numbers – numbers less than zero

net – a 2-D shape that can be folded up to make a 3-D shape

numerator – the top number of a fraction

p.m. – short for 'post meridiem' meaning 'after noon', used to show times between 12 noon and 12 midnight

partition – to break numbers down into units, tens, hundreds and thousands

perimeter – the distance around the outside of a shape, measured in cm, m or km

polygon – a 2-D shape with straight sides

positive numbers – numbers greater than zero

product – the answer when two or more numbers are multiplied together, e.g. the product of 4 and 8 is 32

quadrilateral – a polygon with four sides

reflection – the mirror image of a shape

regular polygon – a 2-D shape with all sides and angles the same

remainder – the number left over after division, e.g. 7 ÷ 3 = 2 r 1

right angle – a quarter turn measured as an angle of 90°

round up/down – writing a number as an approximate, e.g. 64 rounded to the nearest ten is 60

sequence – a set of numbers written in an order following a rule, e.g. 1, 4, 7, 10 is a sequence adding 3 each time

tetrahedron – a 3-D shape with four faces that are all triangles

translation – moving a shape by sliding it to a different position up, down or across without turning it

Venn diagram – used for sorting things into sets

vertical – a line that points straight up at right angles to a horizontal line, a line parallel to the sides when represented on paper

vertex (vertices) – the corner of a shape, where sides or straight edges meet